Adeline Cheang

S0-ARN-817

Sold

DEVELOPMENTS IN FOOD PRESERVATION—1

THE DEVELOPMENTS SERIES

Developments in many fields of science and technology occur at such a pace that frequently there is a long delay before information about them becomes available and usually it is inconveniently scattered among several journals.

Developments Series books overcome these disadvantages by bringing together within one cover papers dealing with the latest trends and developments in a specific field of study and publishing them within *six months* of their being written.

Many subjects are covered by the series including food science and technology, polymer science, civil and public health engineering, pressure vessels, composite materials, concrete, building science, petroleum technology, geology, etc.

Information on other titles in the series will gladly be sent on application to the publisher.

DEVELOPMENTS IN FOOD PRESERVATION—1

Edited by

STUART THORNE

Department of Food Science and Nutrition,
Queen Elizabeth College, University of London, UK

APPLIED SCIENCE PUBLISHERS
LONDON and NEW JERSEY

APPLIED SCIENCE PUBLISHERS LTD
Ripple Road, Barking, Essex, England
APPLIED SCIENCE PUBLISHERS, INC.
Englewood, New Jersey 07631, USA

British Library Cataloguing in Publication Data

Developments in food preservation. (The
Developments series)
1.
1. Food—Preservation
I. Thorne, Stuart   II. Series
641.4   TP371.2

ISBN 0-85334-979-7

WITH 37 TABLES AND 99 ILLUSTRATIONS

© APPLIED SCIENCE PUBLISHERS LTD 1981

Printed in Great Britain by Galliard (Printers) Ltd, Great Yarmouth

PREFACE

Unless they are processed to prevent it, most foods will deteriorate rapidly after harvest or slaughter and will cease to be acceptable or safe. This post-harvest wastage—published estimates of its magnitude vary from 10 % to 50 %—is a major contributory factor to world food shortages.

The major cause of wastage in food is the growth of micro-organisms. Physiological changes in the produce and other chemical reactions such as oxidation also contribute to deterioration, comparatively simple reactions often being responsible for nutritional deterioration such as vitamin loss. Most food preservation processes are effective because they create an environment hostile to microbial growth; low water activity, low temperature, high concentrations of solutes, low pH or combinations of these. Canning and related processes alone rely on a rather different principle, that of actually destroying micro-organisms in foods and preventing their subsequent ingress. It is fortunate that processes that inhibit microbial growth also slow the rate of most other deteriorative reactions.

The aim of any preservation process must be to eliminate microbial and toxicological hazards whilst maintaining the nutritional and sensory attributes of the food. Processes are usually designed to effect microbial safety and stability. Maintenance of sensory quality is usually adequate, since the processor is well aware that his customer will reject products of poor sensory quality. However, nutritional quality and the effects of processing on this are often neglected; it is not an obvious characteristic of food and, in Europe and North America at least, moderate nutritional deterioration may not be of immediate consequence to most of the population. But it is the very old and the very young, both of whom are most susceptible to nutritional deficiency, who rely most heavily on processed foods. It is rare for a preservation process to be optimised for both adequate stability and minimum nutritional deterioration. But the situation is improving slowly. High temperature heat processes, such as aseptic canning, were introduced because of the improved sensory quality of the products and for technical reasons, but a secondary benefit of these processes is reduced nutritional deterioration. Energy considerations favour the replacement of thermal evaporation by membrane processes for the concentration of some products; again there is usually a reduction in nutrient destruction.

No process can endow the product with an infinite life; it is never possible to eliminate completely every deteriorative reaction. But some come close to it. Century-old cans of meat have proved to be harmless if not too palatable! It is not usually necessary, however, to achieve very long storage life and storage in excess of 12 months is rarely needed. If the storage life is very long it is usually better to consume the product early in its life when its quality will be better. With many processes, a compromise must be effected between cost and storage life. The type of packaging and storage conditions, in particular, can have a very considerable effect on both storage life and on costs. When considering this compromise, it must be remembered that deterioration is a continuous and gradual process; processed foods rarely show a sudden transition from completely acceptable to completely unacceptable after a definite and well-defined time. When considering the compromise between cost and storage life, it must be remembered that a product with a total storage life of, say, 12 months will be in a much better condition after six months than a product with a total storage life of six months, which will be only just acceptable.

Until recently, little attention was paid to the energy requirements of food processing and preservation. It was often maintained that, since the value added to food by the processing itself was very small compared with the value of the raw materials, economies in the energy consumption of the processing were inconsequential. But increasing energy costs have made energy considerations a major field of interest in recent years. I consider that such studies will become even more important during the next decade. Investigations into reverse osmosis as an alternative to thermal evaporation and improvements in the thermal efficiency of canning plant are two areas where reduced energy consumptions are being achieved; both are discussed in this volume.

Almost all foodstuffs are stored in a fresh state, even if they are later to be preserved by another process. Fresh storage alone is still the major food preservation process and I believe that its typically low energy consumption will increase its importance in the future. The major disadvantages of fresh storage and distribution, from the food technologist's standpoint, are the limits that it imposes on storage life, the difficulty of predicting this storage life and lack of control over distribution conditions. A great deal of attention is being paid to the storage behaviour of fresh produce and it is now possible to predict storage behaviour with some precision, though not always yet with adequate precision. Recent advances in our knowledge of the storage behaviour of fresh produce are reported in two reviews in this volume. Even when fresh storage is but a preliminary stage of some other preservation process, adequate attention must be paid to it, for quality lost cannot be regained.

It is in developing countries that the greatest post-harvest and post-slaughter losses of foodstuffs occur; and it is these countries that can least afford such losses. They can be reduced by the introduction of appropriate food preservation technology. But the technology that is often introduced is western technology, with large, centralised processing plants of great complexity. Such plants are often built through lack of knowledge of local conditions and needs and because they confer prestige. Because the needs of developing countries should be a major concern of food technologists, it is appropriate that the first chapter in this volume deals with the technology that is appropriate to and fills the actual needs of developing countries.

Food preservation is more than engineering and its proper understanding must involve the relationships between engineering, nutritional, microbiological, biochemical and economic aspects of preservation. Only the amalgam of these can be considered to encompass food preservation as a whole. It is, of course, necessary to specialise in some aspect of food preservation and contributions in this series will reflect this need for specialisation. But it is the aim of this series to present monographs on all aspects of food preservation. It is hoped that the wide range of topics will help practitioners of food preservation to keep abreast of developments in their own field and to appreciate the significance and limitations of aspects of food preservation other than their own.

STUART THORNE

CONTENTS

LIST OF CONTRIBUTORS

José Segurajauregui Alvarez
Department of Food Science and Nutrition, Queen Elizabeth College, University of London, Campden Hill Road, London W8 7AH, UK. Present address: Abedules 69, Jardines San Mateo, Naucalpan, Mexico.

Marc R. Bachmann
Department of Dairy Science, Swiss Federal Institute of Technology, Eisgasse 8, CH-8004 Zurich, Switzerland.

G. van Beek
Sprenger Instituut, PO Box 17, Haagsteeg 6, Wageningen, The Netherlands.

M. Demeczky
Central Food Research Institute, Budapest II, Herman Ottó út 15, Hungary.

E. Godek-Kerék
Central Food Research Institute, Budapest II, Herman Ottó út 15, Hungary.

Margaret A. Hill
Department of Food Science and Nutrition, Queen Elizabeth College, University of London, Campden Hill Road, London W8 7AH, UK.

M. KHELL-WICKLEIN

Central Food Research Institute, Budapest II, Herman Ottó út 15, Hungary.

J. LORENTZEN

A/S Atlas, Baltorpvej 154, DK-2750 Ballerup, Copenhagen, Denmark.

H. F. TH. MEFFERT

Sprenger Instituut, PO Box 17, Haagsteeg 6, Wageningen, The Netherlands.

JUHANI OLKKU

Food Research Laboratory, Technical Research Centre of Finland, Biologinkuja 1, SF-02150 Espoo 15, Finland.

STUART THORNE

Department of Food Science and Nutrition, Queen Elizabeth College, University of London, Campden Hill Road, London W8 7AH, UK.

L. W. WILLENBORG

Stork Amsterdam B.V., Ketelstraat 2, PO Box 3007, 1003 AA Amsterdam, The Netherlands.

Chapter 1

TECHNOLOGY APPROPRIATE TO FOOD PRESERVATION IN DEVELOPING COUNTRIES

Marc R. Bachmann

*Swiss Federal Institute of Technology,
Zürich, Switzerland*

SUMMARY

There are several reasons for not using the food preservation techniques of industrialised countries in the Third World. Western methods usually involve large, centralised processing plants which are usually too far from areas of production for rural transport to cope with. And they are technologically too complex to be constructed or maintained locally.

This review considers those basic principles of food preservation that are of particular importance in developing countries and which allow methods and equipment to be used which exactly fit local needs. A detailed description of a dairy products processing plant, designed with special regard to its location, energy supply and technical equipment, is given. Standards and quality control procedures for use in developing countries are discussed and attention is given to the training of personnel for food processing and preservation operations.

1. INTRODUCTION

The president of the Club of Rome, Aurelio Peccei, says: 'Man has created his own man-made world, where everything interacts with everything else at ever higher levels of complexity, thereby radically altering the natural flows of life on the planet, including his own life. But culturally and behaviourally he is lagging behind. A "human gap" is thus created, which endangers his

1

existence, no less than the mutations in the environment endanger the existence of every species which is unable to adjust to them'.[1]

In these apt words Aurelio Peccei describes the contemporary situation in the industrial countries. Peccei is not the only one to express doubts on the quality of industrial development. In his book *Ein Planet wird geplündert* (*A Planet is Being Plundered*)[2] Herbert Gruhl clearly indicates the danger of such a development to the afflicted nations and supports his findings with numerous statements by internationally recognised researchers. It is thus necessary to state that industrial development has now reached and is pushing against limits, and is endangering the existence of human society. These limits are on the one hand of an ecological nature and can be measured against the diminishing resources of raw materials and cultivable land.[3] On the other hand they are innate in man himself and prevent him from being mentally and emotionally in balance with the industrial world he has built around himself.

What does this realisation mean for the future of the developing world? What consequences for co-operative work have to be drawn from it? This chapter is an endeavour to answer these questions. However, we limit ourselves to the restricted but important field of food technology. We shall attempt to illustrate the theoretical considerations with examples of personal experiences and demonstrations arising out of 20 years co-operation with developing countries.

Before we can deal with the subject it is necessary to make two statements. The first applies to the term 'Third World'. The 'Third World' is as far from being a unity as the 'First World'. To sum up the many different developing countries with the collective expression 'Third World' is a continuous source of misunderstanding and misinterpretation. The differences and the possibilities of development vary considerably from country to country and from region to region. General solutions are as meaningless for development work as global judgements. On the other hand we cannot ignore the fact that developing countries have certain problems in common. To these mutual problems belong, for example, the periodic foodstuff shortages, the modest standard of training amongst the people and the lack of sufficient jobs and money-earning opportunities. If we, notwithstanding the above, use the term 'Third World' in the following it is because we wish to stress common factors which are no respecters of territorial boundaries.

One final remark refers to the term 'food technology'. By food technology we mean all the methods employed to render food imperishable, negotiable and eatable. The term 'food technology' is therefore superior to the term 'food preservation'. 'Food preservation' is a part—the most important part—of food technology. In this present work, therefore, both terms may be understood as synonyms.

2. DOES THE THIRD WORLD NEED A THIRD WAY FOR FOOD PROCESSING?

The initial question posed above as to whether other methods must be employed to preserve and process food in developing countries than in industrial countries can be answered only after the consideration of three further questions. These are: 'Is food technology important for the economic development of a country?', 'What is the present standard of food preservation and processing in developing countries?' and 'What are the specific requirements demanded by food technology in the developing countries?' Whoever attains clarity over these three questions is in a position to judge whether or not the methods of preserving and processing food as practised today in the industrial countries pertain to the developing countries also.

2.1. The Role of Food Technology in Promoting Development

Food holds a key position in the development of a country. It is the basis of existence of the population and in many cases also the only exchangeable goods. The more self-sufficient a country is in food, the less is its vulnerability and the greater its independence. What decisive influence on the development of a country the food supply plays was clearly evidenced by the ever recurring famines in India, the catastrophic hunger in Sahel during the first half of the 1970s and the critical food supply situation in East Africa today. Not only are enormous sums of money from the national budget necessitated for the import and distribution of provisions, but the economic, social and political structures threaten to collapse under the weight of supply difficulties. Examples illustrating this danger were provided by Ethiopia, Chad and Central Africa following the Sahel catastrophe. Development efforts of many years' standing can be wiped out by the lack of food. Even if international help of food is given the sufferings of the larger part of the population before and after a food shortage cannot be ignored. These are caused by the recurring difficulties in supply and rising costs usually due to the black market.

Foodstuffs, stimulants such as tea and coffee, as well as vegetable and animal fibres are for many developing countries the most important raw materials. The ability of these countries to build up capital depends on the production of these raw materials. Only in a few privileged developing countries do oil and other mineral resources represent an additional source of income.

Food and fibres are, however, not only extremely important commercial products, they are also the foundation for local crafts. The first technical capabilities developed by man concerned the provision and preparation of food, clothing and housing. All over the world man has learnt very well to provide himself with the basic goods for existence. In every Third World country we come across there are very typical methods of preparing food and characteristic ways of using fibres and skins for dressing. These original technologies are very well adapted to the local conditions. The local food and other raw materials are used in an appropriate and above all in an economic way. Food technology is not only part of the culture values of a nation but is also the basis for various trades and occupations. Milling grains, pressing oil, baking bread, producing meat, processing milk, tanning skins and furs, spinning and sewing, etc. are among the first professional activities of man. Many prospering industries of the western world have grown out of small food processing units or tiny workshops making tools for transforming and preparing food. Some of the richest and biggest companies in the world either deal in food, or process it. Others fabricate packaging material, and yet others make transport and processing equipment for food. This indicates that food and food processing are very well capable of promoting development. Food technology can be applied with every degree of labour and capital intensity. It can be adapted to any local conditions and to any level of development. If food processing is set up in rural areas, it offers additional working and earning possibilities to that population. It can in this way counteract the rural exodus and subsequent pauperisation of millions of people in the urban slums. This shows again that food technology is a very powerful instrument of development.

2.2. What is the Present Standard of Food Preservation and Processing in the Developing Countries?

The original food technologists all over the world have been (and still are) the housewives. In subsistence economies, the families normally prepare their food directly from their cultivated plants and from their animals. Slaughtering animals and milling grains are done in or around the living quarters. Preservation, which is one of the main objectives of food technology, is achieved by simple means like drying, salting, fermenting and smoking. These preservation techniques as well as the preparation and combination of different foods answer the specific need of the local population in their own environment. It is known that food losses are usually small as long as traditional preservation methods are used.[4] It is also known, that traditional societies use different food according to seasons. Food and human beings form an equilibrate system, at least as long as food scarcity does not exist.[5]

However, with the increase of population and with a growing interdependence of people of different areas, a division of labour takes place. Consequently the food technology, which originally was simply part of the primitive home economy, becomes a separate discipline. Small flour mills, oil mills, butcheries and bakeries are set up. Cooked and ready-made food is sold, either at home or on the market. This is how food technology and catering come alive. As long as food transformation takes place on such a small-scale industry basis, it enjoys a relatively high independence and is not subject to great economic risks. Simple economic structures are still functioning, e.g. one pays for milling the grain by leaving a small part of the flour for the miller. Such a small-scale food industry is well adjusted to the world in which it takes place. It uses local resources, depends on locally made equipment and respects local customs and traditions. In the western world, a more centralised, labour extensive but capital intensive food industry grew out of the cottage-type food industry. It became more and more mechanised because labour was the heaviest cost factor. By eliminating one single labourer, the capital costs of a large machine could be met.

In developing countries, unfortunately, the system of the locally grown food technology breaks down as soon as it leaves the stage of cottage industry. The modern food industry of developing countries is not a continuation of what existed before, adapted to new circumstances. It is rather a strange copy of techniques and systems which have been developed abroad under different conditions and for different aims. Home breweries disappear and are replaced by big, steaming plants producing 'Pilsner', mainly with the help of imported raw materials. Nice smelling local bakeries baking round, flat bread in wood-heated ovens, as for instance in Afghanistan and Pakistan, are replaced by factories producing absolutely white and tasteless English bread. Fruits are not sun-dried any more, but canned in a huge, noisy plant, according to the standards of a multinational company. These are just a few of many examples.

Why this break of tradition? Why this break in continuous and adapted technological development? Is there only one modern food technology, the western one? Hopefully, this is not so. For the sake of many developing countries, we hope that there is an alternative to the extravagant and wasteful western food technology. We shall see later that neither the energy nor other resources would be available for the western type food technology in many countries. Furthermore, it would be difficult to understand why a technology which has been designed to replace labour by capital investments should be applied in countries where labour is abundant and cheap and where unemployment as well as the lack of capital are the greatest problems. However, before the weak spots of the so-called 'western' food technology are pointed out, the question why modern food technology is usually a western food technology remains to be answered. On the one hand, western techniques are indiscriminately taken as the standard of development. Foreign technicians are asked to build up systems and promote techniques which they have learnt in their own countries. Foreign experts who have only a limited insight into the problems and prevailing facts of life of their host countries, quite naturally tend to opt for systems and techniques with which they are familiar. Their usual advice is to follow the same pattern as in their own home country.

This is one answer to our question. The second one lies in the fact that initially the modern food industry of many developing countries is mainly export orientated. The food industry is developed for a foreign, a western market, and not for the home market. The products therefore have to meet the requirements of foreign standards.

However, there is yet a third reason why western techniques are preferred—management. It is certainly easier to manage machines than a great number of labourers.

The fourth and last reason for setting up 'western' food factories in the Third World is bound up with the activity of the bilateral or multilateral developing organisations. Their financial aid in fitting out production plants often leads to dictating, or at the least, advising on the source of equipment. In addition the developing countries often cannot resist the temptation not to take the foreign exchange expenditures for equipment, spare-parts and know-how fully into account, at least not in the initial stages when such installations are financed by the afore-mentioned foreign aid programmes. However, after a couple of years, when foreign know-how is withdrawn and when the need arises to replace the equipment, the dependency on foreign exchange and on foreign countries becomes evident.

In summing up it must be said that at the present time in practically every developing country, traditional food technology and its accompanying skills are on the verge of being lost. Taking their place are imported, inappropriate technology, imported 'know-how' and inappropriate new products.

2.3. What are the Specific Requirements Demanded of Food Technology in the Developing Countries?

In point of fact there exists only one single requirement in the choice of technology—namely to be suited to its location and the conditions prevailing there.

First of all, the customs and traditions of the country should be taken into consideration, in other words, the technology has to be adapted to the people using it.

The technology has also to be adapted to the local resources and these resources have to be used as carefully and as economically as possible. Galtung[6] speaks of 'self-reliance' in connection with the future of technology transfer. An adapted technology, therefore, is based on the use of local raw materials, local manpower, and wherever possible, locally available energy sources. By using locally available manpower, the technology has to be adapted to the level of training and to the existing possibilities of further training. Such a technology will simultaneously help to solve another prevailing problem: the creation of working opportunities. Gibbons[7] pointed out that in Kenya the most severe obstacle for the further increase of food industry is the lack of well trained labour, medium and higher staff. It is pointless to wait with the development of food industry for one or two generations, until there is enough trained staff. It is more reasonable to adapt the techniques to the present educational situation and adapt them again in the future.

The equipment and machinery used should also be adapted to the local conditions. If, for example, there does not exist a tight network of maintenance and repair services, the construction of the equipment and machinery should be as simple and solid as possible. Local material should be used and local manufacturing techniques applied, in order to make the construction of the equipment in the developing country possible and in order to reduce the necessity of importing spare parts to a minimum. In this way, the development of one sector of the national economy can contribute to the development of another, the metal and wood-working sector.

Unfortunately, the situation in developing countries is quite different at present. Agricultural development, including the food industry, rarely leads to a synchronised development of the local machine industry. It rather helps to develop the machine industry of the industrialised countries.

An adapted food technology also makes sure that not only raw materials, equipment and manpower are locally available, but also the necessary chemicals, food additives and detergents. That means, that trade has to be developed for those items which are not available in the country. Local traders have to be encouraged to keep reasonable stocks of the required materials. Such trade asks for the assistance of governments, a fact that leads us to the last aspect of our problem.

An adapted technique needs the support of adapted legal rules and regulations. It is not the role of governments to run food industries—as unfortunately is to be found in many developing countries—but to assist private initiative to run such industries, by introducing well-conceived regulations and standards.[8] It is no use adapting the food technology to the local situation and at the same time being subjected to regulations that have been made in other countries for different conditions or which date from colonial times.

The aim of every adapted technology is not only to omit new dependencies, but also to reduce the existing ones.[9]

2.4. Does the Food Technology Applicable to the Industrial Countries come up to the Requirements of the Third World?

One may assume from the preceding sections that food technology suitable for the industrial countries is not suitable for many countries of the Third World. The assumption becomes assurance if we examine some of the aspects of western food technology more closely. What is particularly telling in this respect is the energy required for the foodstuff system of the industrial countries. This is demonstrated by quoting figures compiled by Steinhart and Steinhart.[10]

Table 1 shows that from 1940 to 1970 the energy input into the United States food system increased on a farm level from 124 units to 526 units, on a processing industry level from 286 units to 842 units and on a commercial and home level from 275 units to 804 units. The total energy use in the United States food system in 1970 was three times as high as in 1940. The energy use in the food processing industry in 1970 was also 300% of the 1940 value.

Figure 1 shows that in 1910, an energy input comparable to roughly 1 cal was needed for the production of 1 food cal. In 1970, already 9 cal of energy were spent in order to produce 1 cal of food for actual consumption. This upward trend is unfortunately continuing. It is well understood that food transformation and preservation at industrial and home level accounts for only half the energy needed. Nevertheless, these sectors also contribute heavily to the ever increasing demand for energy. These figures from the USA appear to us to be representative of most industrial countries.

If we look closer into the energy requirements of different processing and packaging systems[11] we can see that modern processing techniques such as

TABLE 1
ENERGY USE IN THE UNITED STATES FOOD SYSTEM (VALUES TO BE MULTIPLIED BY 10^{12} kcal)[a]

Component	1940	1950	1970
	On farm		
Fuel (direct use)	70·0	158·0	232·0
Electricity	0·7	32·9	63·8
Fertiliser	12·4	24·0	94·0
Agricultural steel	1·6	2·7	2·0
Farm machinery	9·0	30·0	80·0
Tractors	12·8	30·8	19·3
Irrigation	18·0	25·0	35·0
Subtotal	124·5	303·4	526·1
	Processing industry		
Food processing industry	147·0	192·0	308·0
Food processing machinery	0·7	5·0	6·0
Paper packaging	8·5	17·0	38·0
Glass containers	14·0	26·0	47·0
Steel cans and aluminium	38·0	62·0	122·0
Transport (fuel)	49·6	102·0	246·9
Trucks and trailers (manufact.)	28·0	49·5	74·0
Subtotal	285·8	453·5	841·9
	Commercial and home		
Commercial refrigeration and cooking	121·0	150·0	263·0
Refrigeration machinery (home and commercial)	10·0	25·0	61·0
Home refrigeration and cooking	144·2	202·3	480·0
Subtotal	275·2	377·3	804·0
Grand total	685·5	1 134·2	2 172·0

[a] Source: Steinhart and Steinhart.[10]

microwave drying or freeze drying, are extremely energy-consuming. When comparing the different packaging systems, we find that the most energy-intensive system is canning. Ordinary freezing or refrigerating are on the processing level relatively small energy consumers. On the other hand, the same techniques are extremely energy consuming at the transport and domestic level. It can well be said that the western food system is extremely energy-intensive and very demanding in respect of non-renewable resources. Many of the western food technologies must therefore be called

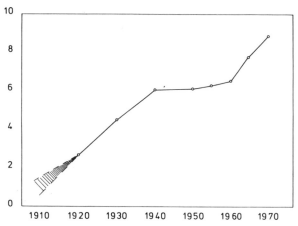

FIG. 1.　Energy subsidy to the US food system to obtain 1 food cal.[10]

TABLE 2

SOME CHARACTERISTICS OF THE FOOD TECHNOLOGY OF INDUSTRIAL COUNTRIES
COMPARED WITH THE RESPECTIVE REQUIREMENTS OF THE THIRD WORLD

Characteristics of industrial food technology	*Food technology as required by Third World countries*
Centralised	Decentralised, labour intensive factories in rural areas
Mainly machinery, labour saving	Creation of many working opportunities
Employment of few, but highly qualified workers	Employment of many poorly and non-qualified workers
Capital intensive	Requiring minimum or no foreign exchange
Use of oil and electricity	Use of locally available, decentralised energy sources
Incomplete use of raw material, much waste	Maximum use of raw material, little waste
High degree of refining, high purity of products	Making use of every available food value
Great convenience, high costs	As inexpensive as possible
Expensive, one-way packaging	Economic return packaging
Internationally and nationally standardised products	Making use of local products, local tastes and local production methods

'unadapted' in respect of the world's natural resources. They will have to be changed in the future and so cannot be genuine and lasting standards.

Besides the high requirements of non-renewable energy and of other raw materials western food technology possesses a whole succession of further characteristics totally unsuited to the needs of the developing countries. These characteristics are listed in Table 2.

When a comparison is made between these characteristics of industrial food technologies and the respective requirements of the Third World, it becomes evident how little suited this technology is to cover the needs of the developing countries.

Modern industrial food technology may well be described as 'inappropriate'. Singer says in this connection: 'Inappropriate technology expressed in inappropriate products prevents employment and raises profit-rates and incomes of the small elite groups (and foreign investors) associated with the modern sector, thus generating income inequalities.'[12]

However, this is exactly what should not happen in a field which is so suitable for the provision of labour and earning opportunities as food technology. A third way is necessary for food technology in the Third World.

3. A MODEL OF AN APPROPRIATE FOOD PROCESSING PLANT

The following arguments are based on many years of experience in building and running milk processing plants in developing countries. Much of the experience thus gained could, however, be equally well applied to other food processing plants. The milk processing plants furnish a decidedly good model case because, compared with other branches of food processing plants, they have to deal with a highly perishable and demanding product. Taking milk as an example it is particularly clear that in warm lands preservation is far and away the most important problem facing food processing. In the case of milk, meat and many vegetable raw materials, preservation (that is the changeover into a stable condition for a short or longer period) is the absolute basic requirement for marketing the goods. Even the simplest form of food processing, namely the grading according to freshness and purity, even if not followed by any other process, is first and foremost a means of safeguarding the durability with an eye to its subsequent marketing.

As well as preserving food stuffs, processing has in many cases the additional purpose of separating the valuable substances in the raw materials from the ballast (i.e. roughage, water, non-participating substances). In the case of milk, for example, full cream cheese represents a concentrate of the most important substances, protein, fat and minerals. This concentrate weighs only about a tenth of the original raw material. In the case of plant raw materials abounding in water, such as fruit, berries, vegetables and so on, drying combines increased durability almost ideally with reduction of ballast. Our reason for discussing both these aspects of food processing is that they are of paramount importance for developing countries. As these countries are also warm or even hot countries, the deterioration of all foodstuffs for chemical and biological reasons is quicker than in temperate or of course cool regions. Therefore more attention has to be paid to the quick preserving of food than in the industrial countries of the moderate zones. The concentration of raw materials to their main component parts is of far more consequence for developing countries than for industrial ones. Developing countries very often lack a good and speedy network of communication. This results in transport being slow, unreliable and expensive. It is therefore obvious that no ballast material but really only high quality integral substances (in a preserved form) should be transported. Food technologists not familiar with the conditions prevailing in developing countries run the risk of underestimating these basic differences to the conditions in industrial countries. That is the only explanation for systems from industrial countries repeatedly being taken over and copied in a seemingly thoughtless manner.

In the discussion on a model of an 'appropriate food processing plant' we propose talking first about the most suitable site, then size and mechanical equipment and finally power supply. We must not forget that 'appropriate' means whatever is best suited to the local conditions.

3.1. Site of the Food Processing Plant

From our earlier remarks the fact emerges that the easy transportability of food in the developing countries is of great significance. Whenever possible food is to be transported over longer distances in a durable and concentrated form, i.e. partly or completely processed. The logical result of this principle is that the processing plant must lie in the production area itself. This is the exact opposite of what one frequently observes in the industrial as well as in the developing countries. Very often food processing plants are situated in or near urban agglomerations. The raw materials for these plants often have to be transported over long distances. Particularly for easily perishable raw materials which are rich in ballast material (e.g.

milk and fruit), such a centralised processing system is not practical. The following rule for the siting of food processing plants in developing countries should be observed:

The more perishable a raw material is and the greater the ratio of ballast it contains, the shorter the distance between the producer and processing plant must be.

This means, for example, that for milk, which is extremely perishable and whose content of water is around 87%, the delivery distance to the processing plant must be kept very short. There is usually no point in giving such distances in miles. Of importance is the time it takes to get the raw material to the processing plant without loss of quality. For fresh milk the radius of the collecting area is around 2 h distance, i.e. the product has to be in the processing plant within 2 h of milking. For less perishable products and those containing less ballast the radius may be extended to 6 or 12 h distance. In setting a maximum total number of hours distance of the collecting radius the varying quality of the transport routes and conveyances have automatically been taken into account. It does not matter whether the raw materials are transported for the named length of time on foot, by ship or by car. In the developing countries the distances over which raw materials are taken to the food processing plants are usually short, as in many cases animals or people provide the most suitable mode of transport. Therefore in contrast to the conditions prevailing in industrial countries, food processing in developing countries must be decentralised.

The decentralisation of food processing has its disadvantages, but also great advantages. The main disadvantage lies in the fact that in the rural production areas the infrastructure for industrialisation is generally lacking. The settlements in these areas only rarely possess an extensive water supply, a reliable source of energy or an adequate network of communications. When setting up rural food processing plants one must take these marginal restrictions into account. The food processing plant must be specially built for its rural location.

A further difficulty is that for such rural food processing plants practically no examples or models can be found. The small cheese dairies of the alpine countries Switzerland and Austria could represent the rare exceptions. This means for the developing countries that there are as a rule no models for them to copy from the industrial countries, so they have to plan and build their own rural food processing plants in keeping with local realities. There can be no more question of purchasing western technology in these cases. The plants must originate on the spot.

Now the advantages of rurally situated plants must be indicated, too. These are partly of an economic and partly of a social nature.

Economic advantages arise from various sources. First, it costs the manufacturer nothing to have the raw materials delivered at the processing plant. The producers can see to the transport themselves because of the short distances involved. The product is payable at the factory. Second, the manufacturer has fresher raw materials to deal with than if he himself had to transport them a long way from a rural collecting centre to the processing plant. Also, the manufacturer can pay for the goods according to their freshness upon delivery. For raw materials which are no longer suitable for the manufacture of high-quality products he pays less than for fresh ware. In the case of milk, for example, this means that liquid milk over a certain degree of acidity at delivery fetches only half or less of the normal price. This is justified, as such milk lends itself very poorly to processing.

In many developing countries they resort to the help of artificial refrigeration when collecting easily perishable goods. For milk they set up in the villages 'chilling centres' which serve as collecting stations. From there the cooled product is transported over long distances to the processing plant. However, we know from experience, that such a system is very liable to breakdowns and also very expensive. To produce cold areas under tropical conditions and to transport cooled milk or other perishable goods once or twice a day over long distances is indeed extremely costly. Furthermore, the chilling of the unprocessed goods does not yet make them ready for consumption, it only makes them more expensive. Neither can one improve the quality of the products by chilling. The best we can hope for is that the quality remains the same as it was at the collection centre.

The transport of preserved food is a different matter. Normally such products do not require to be transported daily to the collection centre but only once or twice a week. Frequently, too, the processed product needs no chilling for transport. The omission of refrigeration during transport and the less frequent journeys are very important economic advantages of the rural food processing plant. Besides this, attention must be given to the risks affecting all transport in developing countries. Bad roads and insufficient maintenance of rolling stock and motors are often the cause of disruptions and delays. In these cases, due to their better keeping properties, processed foods deteriorate far less quickly than the unprocessed raw material.

Rural food processing plants create jobs and earning opportunities and this in turn brings direct social and economic advantages. This is especially essential in developing areas of dense population. Normally such areas are short of arable land. If here more opportunities for earning are not provided the result is poverty and migration to the cities. Migration to the cities and uncontrolled growth—that is to say slums—are at the root of the greatest social problems in many developing countries.

Industrialisation must not be a supplementary reason for attracting more rural inhabitants to the towns. The rural population must be retained in its ancestral surroundings with all the means at our disposal. The fundamental requirement here is the creation of additional earning opportunities in rural areas. What could be a more appropriate way of achieving this than the setting up of a decentralised small-scale food industry?

3.2. Size and Equipment of the Rural Food Processing Plant
3.2.1. The Size of the Plant
The size of the rural food processing plant is on the one hand defined by the amount of raw material produced within its reach and on the other on how complicated the processing is. The larger the production of raw material near and around the plant and the simpler the process, the bigger the processing plant can be. The limitation on size imposed by the density of cultivation or production within the collecting radius (as discussed in the preceding section) is obvious and so we can dismiss that. The second limiting factor, the complexity of the processing, requires further definition. In simple working processes not every step has to be carried out by skilled workers, which is, however, necessary in complex processes. In the case of a rural processing plant, which for obvious reasons may not be highly mechanised or automatic but employs a lot of unskilled or semi-skilled workers, the limited number of well-trained superintending staff (foremen) available, restricts the size of the plant. In milk processing, for instance, it has been proved that about 500 to 3000 kg of milk can be dealt with daily and efficiently under the supervision of one or two well trained specialists. Losses caused by breakdowns in the plant are likewise greater with an increase in the amount of raw material. In order to preserve control and avoid possible losses, small processing plants—but a large number of them—should be set up.

The small plant has further advantages in a developing country. As we shall see in the next section it is easier to make use of unconventional sources of energy such as solar energy, water power and biogas than in a large plant. The small plant is also easier to set up. Small plants can be constructed out of local building materials and using local building methods, which cannot be said for large factories. Finally, the problems of water supply, waste utilisation and sewage are more easily solved in small plants than in large factories. The 'effects of scale' which are well known in the factories of the western world are of a negative nature in the rural plants in developing countries. 'Scaling up' brings no economic advantages under these conditions. On the contrary one could quote many cases to show how large plants built along western lines lead to economic losses. Particularly in the sphere of milk processing there are in many developing countries large plants either showing a deficit or existing only as development ruins and scrap heaps.

3.2.2. The Mechanical Equipment of a Plant

As far as the mechanical equipment of the rural food processing plant is concerned one must be guided by four points of view.

The first concerns the usually insufficient infrastructure in the rural areas, especially the often unreliable supply of energy.

The second point is the necessity of creating as many jobs as possible.

The third aspect one has to consider when equipping a plant with machines and tools is reliability.

The last point—one of far-reaching importance is the impulse towards development which these plants are able to give to the metal and wood manufacturing industry of their own country. All these points force the constructors of food processing plants to act according to existing factors in the widest sense when deciding on mechanical equipment. It will have to be clarified which parts of the manufacturing process it would be sensible to carry out by hand. This manual work, however, must be neither too strenuous nor too monotonous and boring. Experience teaches us that hard physical work and boring work get badly done. The right to work in Article 23 of the General Declaration of Human Rights by the United Nations[13] guarantees every person satisfying working conditions. In the organisation of manual work, therefore, particular attention must be paid to 'satisfying' working conditions. On the other hand all the possibilities of local fabrication of machines and tools will be thoroughly examined and the equipping of the plants arranged in accordance with these possibilities.

These four points lead us to heed the following principles for the mechanical equipping of the plants:

(a)   Tools and equipment are adjusted to the capability and level of training of the local workers.
(b)   The service, maintenance, repair and replacement of the equipment is possible in the country.
(c)   Local energy is used.

(d)   The equipment is inexpensive and is purchased mainly without foreign exchange.

(e)   The cleaning of those parts of the equipment which are in contact with food is easy.

(f)   Apart from models and prototypes, the equipment is constructed in the developing country itself. Local material and traditional techniques in the country of destination are taken into account when constructing prototypes.

Such equipment is usually not available on the market. It is either not manufactured any more or has never been built. We therefore have to invent or reinvent adapted equipment. This can be a very difficult and time-consuming process. Figures 2–5 show examples of such newly developed equipment for rural milk processing. Figures 2 and 3 represent a simple double-jacketed vat with solar heating for cheese and ghee making. Figures 4 and 5 show a hand-operated cheese-melting pan for reprocessing cheese or a cheese-like half product. Further details of this equipment and information on its use have already been published.[14]

3.3. Energy for Rural Food Processing Centres

Not only the processing unit itself, but also its energy sources and the kind of energy used have to be adapted to local conditions. The limitation in size of the rural processing plant opens new and highly desirable possibilities of using locally available and renewable energy sources, which cannot be used by big and highly mechanised factories.

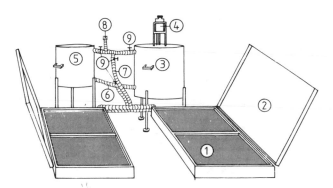

FIG. 2.   Cheese-vat with solar heating. 1, Absorber; 2, covering of absorber; 3, storage tank; 4, expanding tank; 5, cheese-vat; 6, cold water connection; 7, hot water connection; 8, water inlet; 9, valves.

FIG. 3.   Double-jacketed vat with solar heating for cheese and ghee making.

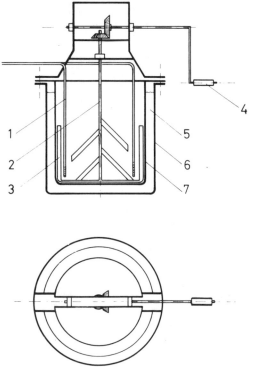

FIG. 4.   Hand operated cheese-melting pan. 1, Steam pipe, steam injection; 2, stirrer; 3, scraper; 4, crank handle; 5, heating water; 6, jacket; 7, melting pan.

FIG. 5.   Hand operated cheese-melting pan.

When planning the energy supply of a processing centre, the following general rules and principles should be observed:

1. With the increasing perishability of the raw material, an increasingly reliable functioning of the energy system of the processing plant must be achieved.

2. One should use preserving and processing methods which are as little energy-intensive as possible. Freezing or preservation by cooling, for example, require a lot of delicate equipment and expensive energy.

3. For the provision of the necessary processing heat, local fuel and (whenever possible) renewable energy sources, such as solar energy, methane gas from the conversion of biomass, wood or waste materials have to be used.

4. Hard manual labour should be avoided. Such work should be done by mechanical means or with the help of motors. However, in order to avoid breakdowns and interruptions in processing, one should also be able to do all these jobs manually in case of energy failures.

5. All transport of raw materials and products within the factory should be achieved using gravity.

6. For many mechanical processes such as milling, pressing, stirring, mixing, etc. one should use as far as possible traditional energy sources, such as wind power, water power and animal draught power.

In the following sections three non-conventional, but renewable sources of energy will be discussed: gravity, solar energy and the energy we gain by conversion of biomass into methane gas.

3.3.1. The Use of Gravity for Transporting Goods Within the Processing Plant

By making use of gravity we want to avoid pumping or other forms of mechanical transport as much as possible. This is because a number of difficulties have to be expected if extensive use of pumps and other mechanical means of transport is made. The breakdown of pumps and other transport elements can stop the process and lead to the loss of large amounts of perishable raw material. Even if the plant has its own electricity supply, such breakdowns may occur. Furthermore, we run a constant risk to hygiene if we use pumps and other complicated means of transport. Cleaning has to be perfect and since chemical inplace cleaning (CIP) is expensive for a small processing plant and dependent on the right chemicals, one has to dismantle pumps and other transport equipment daily for manual cleaning. By doing this, there is a great risk of damaging gaskets and other parts of the installation. We know by experience that the largest number of breakdowns of processing lines take place because of the non-functioning of electrical pumps. Hand-operated pumps are less problematic but, nevertheless, must be easy to clean and to seal.

By arranging the processing lines on floors at different levels, internal transport of goods can be achieved by gravity. At the beginning of the processing line the raw material has to be lifted to the top of the plant. This work should be done by using external slopes. Figure 6 shows a small dairy factory (3000 kg of milk per day) in Afghanistan which has been constructed according to the afore-mentioned ideas. Raw milk is brought over a slope to the weighing machine and to the reception tank which are at the beginning of the processing line. This slope was made with the material excavated for the building.

The personnel which runs such a 'cascade-type' factory will find out that there are certain differences in comparison to conventional one-floor plants. One has to climb more stairs and avoid spraying machines and products with water while washing the floors above. On the other hand, they will also find that the cleaning of the process line is easier and quicker because there are no pumps or other mechanical means of transport.

3.3.2. Use of Solar Energy for Small Rural Plants

In many developing countries there exist favourable conditions for the

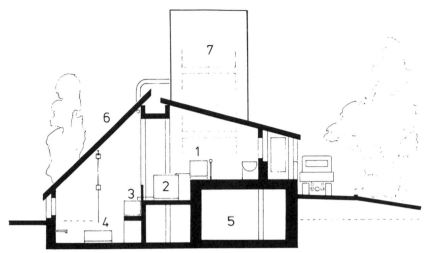

FIG. 6. 'Cascade-type' dairy factory. 1, Reception basin; 2, pasteuriser; 3, balance tank; 4, cheese-vat; 5, cheese ripening room; 6, roof with integrated solar collector; 7, water tower.

installation of solar energy systems. Nevertheless, there are a few points which have to be considered:

1. Solar energy is not constantly available. Even under favourable climatic and geographical conditions a processing plant needs a second energy source apart from solar energy.

2. The less favourable the conditions for the use of solar energy are, the better the stand-by energy systems must be and the more has to be invested for the storing of solar energy.

3. The price of solar energy is equal to the capital costs (including amortisation) of the investments which are necessary for its collection and storing. This price must be compared with the prices of the cheapest conventional energy sources and only if the comparison is favourable, should one start using solar energy. It goes without saying that in today's situation we also have to think of the rising costs of conventional energy and therefore should gain our first experiences with solar energy for food processing while the sun shines, i.e. while we still have time for such experiments.

4. A solar energy system should produce more energy during its lifetime than one needs for the construction of the system. Only if this condition is fulfilled, do we have a positive energy balance, or in other words, a positive 'harvesting factor'.

5. Solar energy systems have to be simple and made in such a way that they can be built in the developing country itself.
6. Before starting to build a solar energy system, one has to know the basic irradiation data of the area. It is worthwhile going to some trouble in finding these data or in recording them over several years. Tables exist for the approximate irradiation data of nearly the whole world. The specific local conditions, however, have to be recorded on the spot.
7. Solar energy should be used for those thermal processes which take place below 90 °C. Temperatures below the boiling point of water are easily achieved with flat plate solar energy collectors. In order to obtain higher temperatures one needs concentrating collectors. These collectors are expensive and difficult to construct and need a very elaborate system for tracking the sun. Their biggest disadvantage, however, is, that they make use of direct irradiation only and cannot use diffuse irradiation. Luckily enough, there are many heating processes for the preservation and treatment of food such as drying, pasteurising, fermenting and so on, which are carried out at temperatures far below 100 °C.

In addition to these general considerations we would now like to go into some technical details of a solar energy system. Whenever possible, flat plate collectors should be part of the roof of the building. Figure 7 shows our small dairy plant in Afghanistan with a 90 m² roof-cum-solar collector. On a simple supporting construction a good layer of insulation material has been fixed. On this insulation material a tube system made out of simple 1 in water tubes is installed. The individual tubes, which have a spacing of 20 in

FIG. 7.   Dairy with solar energy collecting roof.

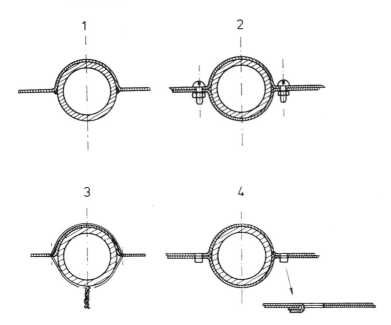

FIG. 8.   Sheet tube constructions for solar collectors. 1, Soldered; 2, screwed; 3, wired; 4, embossed.

are provided with metal sheets which are well fixed to the tubes (Fig. 8). The surface of the metal plates is painted black. The uppermost cover of the roof is made out of solid transparent plastic sheets or fibre-glass sheets. With this type of collector it is possible to collect between 10 a.m. and 3 p.m. 30–40 % of the energy of the solar irradiation and to accumulate it in the form of hot water at temperatures up to 85 °C.

A less sophisticated flat plate collector can be made by fixing the tubing system directly to the black painted corrugated metal sheets of the roof. This very cheap type of flat plate collector should, however, only be made in warm areas and if a temperature of not more than 40–50 °C is needed.

The hot water accumulating system should be as simple as possible and function without pumps. This can be achieved by placing the hot water tank in a higher position than the solar collector (thermosyphon principle, Fig. 9). In regions where the temperature falls below the freezing point, the collector circuit has to be separated from the hot water circuit which is being used for processing. In order to achieve this, a heat exchanger is placed in the hot water tank. Then it is possible to add antifreeze to the collector cycle.

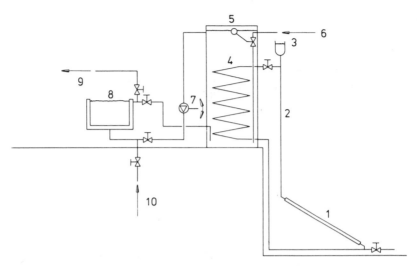

FIG. 9. Thermosyphon principle. 1, Solar collector; 2, thermosyphon circuit; 3, expansion tank; 4, heat exchanger; 5, storage tank; 6, cold water supply; 7, hand pump; 8, cheese-vat; 9, hot process water; 10, cooling water.

From our first experiences with solar energy systems we learnt that:

1. The team which is to build and use such a solar energy system must be convinced of its usefulness. Otherwise they tend to seek proof that this unconventional innovation is no use and one should go over to a conventional system.

2. The whole tubing system of the solar–water circuit has to be absolutely and completely water-tight. Under the conditions in a developing country it is sometimes difficult to perform high quality welding. It is therefore necessary to check all welded parts carefully before using the system. The same applies to all flexible joints, for which only the best rubber tubes and jubilee-clips should be used.

3. One has to avoid heating up the whole amount of solar-heated process water when using a stand-by heat source. Such a heat source should increase the temperature of only that amount of water directly necessary for processing. The stand-by heat source should not raise the temperature of the water in the solar circuit much, because the performance of this circuit decreases with rising temperatures.

In the not too distant future, it will be possible to produce cold also by solar energy. Several groups of scientists are working on this problem. There are already a few prototypes of such sun-activated cooling machines working in different countries.[15-17] However, the main problem still to be solved is, how to make these machines sufficiently simple to build, reliable and easy to operate. A solar cooler is aimed at a machine which uses no other energy source but the sun and which is operated simply by opening or closing a few valves once or twice a day. Until such a machine is available, it would be wise to apply the following principles in connection with the production and use of cold in rural food processing plants:

1. As far as possible natural sources of cold have to be used. Such natural sources are: Running water, deep cellars, storerooms and ripening rooms built at high altitudes, cold from natural evaporation and thermal irradiation to the clear night sky. For many centuries these last two sources of cold have been extensively used in the Near East, India and other countries.
2. Food transformation processes have to be applied which do not need artificially-produced cold.
3. If the use of artificially-produced cold is absolutely necessary, one should try to produce it by means of conventional kerosine refrigerators or by compressors directly linked to reliable diesel or petrol engines. Such equipment has given satisfaction in many developing countries.

3.3.3. Methane Gas as an Energy Source for Rural Food Processing Plants
Bioconversion for energy production is not considered to play an important role in industrialised countries at the present moment.[18] In developing countries, however, bioconversion is one of the most promising means of overcoming the energy crisis. Of all bioconversion systems (methanol and ethanol production, methane gas production, hydrogen production from algae, uranium collection from the sea from seaweed and algae) the methane gas production has the best chances. In spite of the fact that for many decades India and China have accumulated knowledge of biogas production, there was not much technical literature available on biogas utilisation until recently. The books of Meynell[19] and Baader *et al.*[20] are, among others, excellent working aids for the planning and construction of biogas plants.

Not much is known of experiences of the use of methane gas as an energy source for rural food processing plants. Our institute, therefore, in 1978 started a research programme aiming at making a rural dairy factory (5000 kg of milk per day) independent of external energy sources through bioconversion of waste and the slurry of its piggery. In spite of the fact that this research project is taking place in Switzerland, we are convinced that much experience which we hope to gain from the project, can be of use for food processing plants in developing countries.

Before planning a methane production system one has to be clear about the following problems:

1. Methane production is a continuous process. That means that after an initial phase, the biogas plant needs constant and regular feeding with suitable organic material. One has to make sure, therefore, that a sufficient amount of such organic material is always available.

2. A food processing plant which does not have its own piggery or cattle-feeding unit, will usually not have a sufficient amount of organic waste to be converted into biogas. Such plants will have to organise a regular and constant flow of organic material for their bioconversion system. A very promising possibility for warm countries is a combination of sewage treatment with the help of water plants (water hyacinth) and biomass production.

3. For methane gas production the biomass has to be broken up into small particles. If manure is being used, the shredding and breaking up of the biomass has been done by the animal. In all other cases this work must be done by mechanical means. This type of work can be done by hand-operated or animal powered machines.

4. The anaerobic fermentation of biomass does not produce heat. The fermenter, therefore, has to be heated up to the optimal fermentation temperature. By using mesophile methane bacteria this temperature is 30–35 °C. Well insulated fermenters help to keep a constant reaction temperature. The use of solar energy for heating the fermenters is very advisable in warm countries. If part of the produced biogas has to be used for heating the fermenter, the performance of the whole installation is reduced by 10–50 %.

5. The most important parameter for gas production is the dry substance of the organic raw material to be fermented. The dry substance of the slurry which is fed into the fermenter should be within the range of 5–15 %. On the one hand such slurry does not create problems with pumping or mixing, and on the other hand leads to a fairly good gas output. The amount of gas which is produced depends on how long the organic material remains in the fermenter and on the type and composition of the organic material. From 1 kg of organic dry material 400–700 litres of gas can be

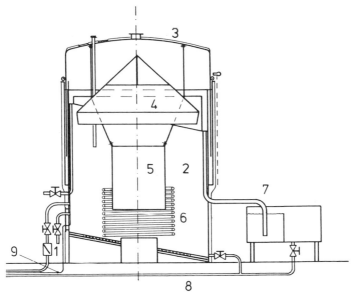

FIG. 10. Fermenter for methane gas production. 1, Inlet; 2, fermenter; 3, gasometer; 4, stirrer; 5, counter weight; 6, heating coil; 7, pressure equalising basin; 8, slurry outlet; 9, gas conduit.

expected. The optimal pH for methane production by anaerobic fermentation is approximately 7. If the material is rich in carbohydrates (cellulose, lignin, starch) the fermentation risks becoming acid and if the material is very rich in proteins, a lot of ammonia nitrogen is produced and the pH gets too alkaline. In order to maintain the optimal pH, the carbon/nitrogen ratio should be between 10 and 16.

6. A special problem is the formation of floating layers of solids from the slurry on the top of the fermenting substrate. These layers tend to become very compact, stopping the gas outlet and plugging even the slurry outlets. The floating layers have to be destroyed by stirring. Stirring can be done without external energy sources by using the gas pressure which builds up in the fermenter. Such types of fermenters have been developed. The fermenters used in the above-mentioned research project are equipped with a conical stirrer fixed to the gasometer of the fermenter (Fig. 10). When gas accumulates in the gas-dome, the conical stirrer is lifted thus breaking up the floating layer and pushing it towards the circular drain of the fermenter.

7. Methane gas production is a complicated fermentation process. The persons running a biogas plant therefore need at least basic microbiological knowledge and have to be fairly well trained for their job.

8. The whole methane gas producing system consists of two main parts: the fermenter and the gas-storing installation. The latter may consist of a gasometer or a simple gas-tight plastic bag or another container. On one side the gas is being continuously produced, on the other side it is being used by the food processing plant only during a few hours per day. Therefore, one has to make sure that there is enough storage for the gas. If seasonal variations in the power demand of the plant exist, the biogas system has to be conceived in such a way that it produces enough energy during the peak seasons.

9. A methane gas system is *not* only producing gas, it is producing a large amount of fermented slurry as well. It has to be clear from the start what should be done with this slurry. It must by no means simply be drained away because it is a highly valuable organic fertiliser.

The above-mentioned problems make it clear that the rural food processing plant needs a very well-planned and well-run methane gas system if it is to be able to rely on this source of energy. However, if it has such a system, it is in a position to be fairly independent of external or imported energy.

The small size of the rural food processing plants and their decentralisation are the secrets in making renewable local energy sources available for food processing in developing countries. This is true not only for the conversion of biomass into methane gas but also for the use of solar energy, wind power, bulky fuel, etc. There is a 'reversed effect of scale' if we look at the energy supply of rural food processing plants. The smaller the plant the simpler becomes its energy supply and the easier is the use of renewable, local energy sources.

4. PRODUCTS TO BE MANUFACTURED IN THE RURAL FOOD PROCESSING PLANTS

It is not the purpose of this section to list selected products and their technological data. The examples are quoted only in order to illustrate the processing under discussion. These examples have again been chosen from the field of dairying.

The principles to be followed in the rural processing plants are:

(a) Adaptation of the products to local conditions.
(b) Production first for the local market and only then for export.
(c) The choice of simple, energy-saving preserving and processing methods.

4.1. Adaptation of the Products to Local Conditions

Not only the means of production such as buildings, machines and tools but also the products themselves are to be adapted to local conditions. Very often food technology which is influenced by the industrial world runs contrary to this principle. It is not the case that it is the products which are adapted to the climatic, technical and economic conditions of the environment, but the other way round. With enormous technical and financial effort refrigerated buildings for the production and storage of western food are being constructed. A parallel case in the industrial countries are the artificially heated hot houses for the production of exotic fruits and vegetables. The rich industrial countries are still in a position to afford such ecological nonsense. It is, however, inexcusable to apply this system to Third World countries. As we have already mentioned, the native food technologies and products are suited to the location. That is true, too, for the industrial countries. Only with the rising claims and wealth among inhabitants of the industrial world came the inversion. Such an inversion is not only expensive, but takes place at the cost of non-renewable resources. It cannot, therefore, last beyond a limited time. Sooner or later we of the industrial countries will have to come back to less wasteful production methods. So it seems idiotic to us to let the developing countries be led 'up a blind alley'.

Suitable products for warm countries are, for example, those that do not need artificially-made cold for their storage and distribution. Suitable products are those that cost little to make and use locally available raw materials. Suitable products are those that retain not only a fraction of the original raw material as food, but make full use of the nutritive value.

An example from milk processing of a suitable product is fermented milk. By the simplest method milk can be made to last for days in a marketable form by spontaneous or induced acidification. As an example of a badly adapted western product pasteurised milk can be quoted. The latter is only comparable to fermented milk from the point of view of keeping qualities if protected by expensive packaging and kept cold the whole way from producer to consumer. It is not at all surprising that in warm countries fermented milk has always been known whereas pasteurised milk has put in an appearance only since the advent of technical co-operation.

Still in the realm of milk processing two more examples can be named. A suitable product is white cheese ripened and kept in brine as it is often to be found in the Near East and in Southern Russia. Unsuited to the conditions of warm developing countries are by contrast European-type soft-cheeses. These products require for their ripening and most particularly for their transport, distribution and storage artificial cold.

In milk processing it has been proved successful to produce low price goods for the low income population as well as luxury products for the well-to-do consumer and for catering establishments. Therefore the raw material, milk, is partly skimmed and the cream thereby gained marketed in the form of fresh butter, double cream and cream cheese. The profit from the sale of these luxury products is then used to reduce the price of the standardised milk and the products made from it.

4.2. Production First for the Local Market and Only Then for Export

It is understandable that the governments of the developing countries want to gain foreign exchange through exports. As complete self-sufficiency or extreme 'self-reliance' for large or small economies is utopian, the village community as well as the country as a whole require cash crops. The importance of export products such as coffee, tea, fruit preserves, etc. should not be underestimated. Despite this the provision of the local population with suitable foodstuffs must not be neglected. These people must be provided with food suited to their conditions at a reasonable price.

This is the principal aim of the rural food processing plant. It is ideally placed for the processing of traditional products such as ghee, koa, ghari and so on using local raw materials.

One advantage of choosing traditional foodstuffs is that the farmers can continue to grow the crops they are accustomed to. They are not forced to switch over to 'modern' crops. They are not forced to grow pineapple instead of maize, wheat instead of millet or soya instead of the traditional dhal. Such a changeover is usually only profitable to the big and rich farmers and not to the small peasants, who have neither the necessary means nor the knowledge. The maintenance of a market for the traditional products is thus a help for the needy small farmer. However, stressing the importance of traditional crops and traditional foods should not put a brake on the development and introduction of high yielding varieties and new crops, but it should help to achieve development without detriment to the needy and poor.

4.3. The Choice of Simple, Energy Saving Preserving and Processing Methods

In Section 4.1. we said that the products should be adapted to local conditions. The same applies to the processing methods. These methods should by no means make the products more expensive than absolutely necessary.

According to Bourne,[21] who studied the problems of food technology in the Philippines, the most appropriate preserving methods for food are drying, preservation by chemical means and canning. From our personal experiences in various countries of Asia, Africa and Latin America, we completely agree with this view. By far the most appropriate means of preservation is sun-drying. This is true in spite of the fact that there is still a lot of work to be done to improve the existing sun-drying systems and the pretreatment of the food to be dried. Above all we will have to develop systems that use warm air for drying and where the food is not exposed to the sun as sunlight can be detrimental to certain substances. Furthermore, it seems that by an appropriate pretreatment of the food, i.e. by salting or by lowering its pH through acidification, drying can be improved.

The chemical preservation methods are traditional as well. Among these we find salting and different types of fermenting. Lactic acid fermentation, alcoholic fermentation and fermentation caused by moulds are classical examples.

The fact that dried and chemically preserved foods can be marketed and stored without cooling, is an enormous advantage of these preserving techniques. Drying combines this advantage with the extremely useful decrease of weight, which makes the goods easier to transport.

Canning offers the advantage of storing at ambient temperatures as well. The big drawback of this method, however, is the high price of the packaging material, tins and glass jars. Unless we can find a way to use returnable tins, as proposed by Bourne,[21] the use of this packaging material is usually not justified for the sale of food on the local market. Providing small rural processing plants with the necessary amount of tin-plate is usually difficult. Only where there is a reliable source of tin-plate in the country itself, can we consider tinning as an appropriate preservation method. In these rare cases, the manufacturing of tins in the rural food processing plant can even help to create additional working and earning opportunities.[22] A further recommendation for the introduction of this preserving method is if the tin is used as a wholesale (returnable) container and not as retail packaging, thus adding little to the price of the goods.

The problem of retail packaging must be carefully studied in every case.

Charpentier and Tuszynski[23] state, correctly, that it is nonsense to add to the price of basic food, such as milk, by selling it in costly one-way packets and thus putting it out of the reach of the low-income population. Returnable packing materials such as glass bottles have a similar effect on costs. Used glass bottles have a considerable market value in a developing country. The customer will therefore return the bottles only if he has to pay a deposit which is higher than the bottles' market value. Such a deposit, however, puts the price of the goods up the first time they are bought or whenever a bottle is lost, rendering them inaccessible for low-income customers.

According to our experience there are two different solutions to the packaging problem. The first one is selling the goods without packaging material directly into the customers' containers. The second is the use of cheap one-way packaging materials such as plastic pouches or grease-proof paper. For the sale of milk both systems have been successfully used. The Mother Dairy in New Delhi maintains a chain of approximately 250 automated milk-selling booths from which the customer buys open milk which he takes back home in his own container. When milk is to be sold in plastic pouches the 'Thimmonier' system is often chosen. Hand operated or semi-automated packaging machines exist for this system.

The packaging problem becomes still more ticklish and difficult when food has to be packed in a country of the warm and humid tropics. In addition to the aforementioned problems the water vapour permeability has to be taken into consideration. Under these conditions and for dried food there is practically no other acceptable packaging material but a well-sealed and, whenever possible, returnable drum.

In summing up we cannot hide the fact that the food packaging problem in developing countries has not yet been solved. Our advice is, rather than look for a general and final solution study each case individually and choose the best possible solution under the prevailing circumstances and keep on improving it.

5.   STANDARDS AND QUALITY CONTROL FOR FOOD IN DEVELOPING COUNTRIES

It is not our intention to discuss here the standards and testing methods for export products. These standards depend in any case on international rules and conventions and can hardly be changed in a day. Furthermore these standards have usually not been made for food processing in developing countries or under tropical conditions, but for the conditions prevailing in industrial countries. They are very often an unfair handicap for food processing under difficult climatic and economic conditions.

We believe that for the rural food processing plant, working mainly for the local market, not these international standards but more appropriate national standards and regulations should be applied. High quality standards are all very well for rich people but not for the average consumer in a developing country. The latter is not in a position to allow himself the luxury of top quality. For him the low price of food is by far the most important 'quality standard'. The standards and quality instructions of the industrial countries are therefore of little value under these circumstances. All the more regrettable it is, that practically all the governments of developing countries act according to rules and regulations which have been made for the industrialised countries, and for a food industry working under completely different conditions. This explains the difficulties which one encounters when trying to apply unconventional methods such as the one mentioned in the previous section for the standardisation of the fat content of milk. It also explains why certain preserving methods, which would be extremely useful in warm countries, such as the peroxide–catalase method[24] are not allowed.

In our opinion the following five standards for food processing rank highest for developing countries:

1. Food has to be free of harmful substances of a chemical or microbiological nature
2. Food has to be free of foreign matter and deliberately added weight
3. Food must be storable for the length of time customary for its normal usage but no longer
4. Food must have the best possible digestability balanced against an economic processing method
5. Food should contain the largest possible amount of its original food value balanced against an economic processing method.

The last two standards are compromises which take the low purchasing power of the population of developing countries into consideration. They make quite clear that we should not aim at top quality with respect to taste and food value at any price.

We have said that a population living at subsistence level cannot afford the luxury of high quality. Many producers and merchants tempted by this make a profit by adulterating their goods. While checking milk of 'milk paddlers' in an Asian country we found milk adulterated with more than 70 % of water. For every honest food processor such fraudulent practices of his competitors are severe handicaps. As long as no system exists for checking and eliminating such sales and unfair competition, normal business is nearly impossible. It would, therefore, be an extremely valuable contribution towards the promotion of food processing, if the governments made rules for the handling of food and at the same time set up organisations to enforce these rules. How to make such rules has been discussed by Siegenthaler.[25] In our opinion it would indeed be much more in the interest of the development of food processing, if the governments concentrated on drawing up rules and regulations for these activities, rather than engaging in food processing themselves.

For quality testing of raw materials at the processing plant or the collection centre, there rarely exists appropriate testing methods. In spite of this, fraud must be prevented from the very beginning of a food processing operation. The most suitable methods are quality tests carried out in the presence of the producers immediately after collecting the goods. Since electricity is seldom available and the continuous supply of chemicals is not guaranteed, the most appropriate test methods are those which depend on neither of them. Under the conditions of rural food processing an exact analysis is not as important as an overall picture of the quality of the goods. For most vegetables and fruits only simple quality tests, e.g. checking for purity, cleanliness, ripeness, etc. have to be performed. For raw materials where the content of dry matter of a specific substance is important, the testing methods become more complicated. Sometimes one needs small presses for obtaining juice or oil and aerometers or refractometers for determining the density and optical properties of these materials. If the acidity of the raw material is a limiting quality factor, one is very often helpless without appropriate titration equipment. Even without such equipment simple tests can often help. There is a general rule that physical testing methods are quicker, simpler and easier for the producer of the raw material to understand than chemical tests. Moreover, for these tests one does not need reagents, which sometimes are difficult to come by and to keep in stock under the conditions prevailing in a developing country. With some inventive imagination it is possible to find appropriate testing methods even for complex raw materials such as milk.[26]

6. TRAINING OF PERSONNEL FOR RURAL FOOD PROCESSING UNITS

Work in a rural food processing plant does not mean switching on and off machines, pressing buttons and controlling automatic operations. Work in a rural factory means really processing food, making the products. The workers in such a plant must get to know and perform many different operations for food transformation. In order to do this they need manual skills. Such skills exist in every rural society and we must take care not to lose them. People working in rural food processing plants must, for example know how to dry coffee beans, how to ferment cacao beans, how to process rice, how to transform milk into butter, ghee and cheese, etc. Such operations cannot be learnt in schools, they have to be learnt the practical way. That is why we are convinced that an apprenticeship of one or several years is the backbone of every training for food processing in developing countries as well as in developed ones.

It is true that setting up a food industry—including small rural food processing plants—needs more than just a practical training. It also needs engineers who know how to draw and construct equipment and machinery for food processing. It needs economists who plan and guide food collection, transformation and marketing. It needs scientists who formulate rules and regulations for food processing and marketing and who can work out appropriate food testing methods. It is true that it needs engineers and architects who know how to build factories. But still we are convinced that the most important persons in rural food processing are those who have learnt their profession by well-guided in-job training. It is a great mistake to imagine that this type of work can be learnt in industrial countries. With the exception of small-scale milk processing, as practised in certain areas of Switzerland and Austria, such a training has to be organised in the developing country itself. As we have already said, one must learn the trade the practical way. That means, that one should not start by building schools or training centres. The wisest thing to do is to select from among the existing food processing units the best ones and to organise courses for trainees in these plants. During such an apprenticeship the trainee has to perform all the operations necessary to make the food and to run the plant, including packing and storing goods and cleaning machines and floors.

In order to make the trainee understand what he is doing, he naturally also needs some theoretical instruction. For this purpose regional courses can be organised where the apprentices learn the basic food processing theories and their chemical, microbiological and technological background. However, such a theoretical training should be closely linked to the practical work. That means, that for each branch of food processing separate courses have to be organised. There should be instruction in milk transformation, in processing fruits and vegetables, in processing cereals and starchy tubers and again in slaughtering animals and processing meat.

Organising such courses is where the duty of the governments should lie, supported perhaps by existing professional societies and institutes.

Following growth of a national food industry, training will also have to be established on a college and university level. Colleges and universities will take over the training of managers for the processing plants, instructors for practical and theoretical courses and specialists for food legislation and control as well as scientists. This higher education in food science can reach its aim only if in addition the trainees of colleges and universities have to spend at least a year gaining practical experience in food processing. If they do not undergo such a practical training, they will never gain sufficient insight into the day-to-day problems of food processing and they will not be able to talk the same language as the workers of the processing plants.

In developing countries the training in food processing should be orientated towards a national rather than a western food technology. Above all it should help to reduce the dependency on foreign industrial countries and encourage a country's own development.

REFERENCES

1. PECCEI, A., *Intern. Development Rev.*, 1978, **XX**, 4–11.
2. GRUHL, H., *Ein Planet wird geplündert*, S. Fischer Verlag GmbH. Frankfurt a/Main, 1978.
3. MEADOWS, D. *et al.*, *Die Grenzen des Wachstums*, Deutsche Verlagsanstalt GmbH, Stuttgart, 1972.
4. REUSSE, E., *Month. Bulletin of Agricultural Economics and Statistics, FAO*, 1976, **25**(9), 1.
5. N'DOYE, T., *Ceres*, 1980, **13**(1), 17–23.
6. GALTUNG, J., *Forum Vereinte Nationen*, 1976, **3**(6), 1.
7. GIBBONS, E. T., *Zeitschrift für ausl. Landwirtschaft*, 1976, **15**, 141.
8. BACHMANN, M., In *Proceedings of XIX International Dairy Congress*, International Dairy Congress Secretariat, New Delhi, 1974, pp. 38–9.
9. MATTHÖFER, H., *Forum Vereinte Nationen*, 1976, **3**(6), 2.
10. STEINHART, J. S. and STEINHART, C. E., Energy use in the U.S. food system. In *Food: Politics, Economics, Nutrition and Research* (Abelson, P. H., Ed.), A.A.A.S., Washington, 1975.
11. OLABODE, H. A., STANDING, C. N. and CHAPMAN, P. A., 'Annual Meeting, Institute of Food Technologists', Anaheim, Cal., 1976.
12. SINGER, H. W., The development outlook for poor countries. Technology is the key. In *Probleme des Technologie-Transfers von Industrie—ländern in Entwicklungsländer*, Sonderreihe angepasste Technologie, Lateinamerikanisches Institut an der Hochschule St. Gallen für Wirtschafts- und Sozialwissenschaften, 1978.
13. *General Declaration of the Human Rights*, United Nations, 10.12.1948.

14. BACHMANN, M. R. and LEUENBERGER, N., *Lebensm.-Wiss.und Technol.*, *LWT-Report*, 1979, **12**, 124–6.
15. HAUSE, R., *Dornier Post*, 1979, **4**.
16. BECHTOFT-NIELSEN, P. and WORSOE-SCHMIDT, P., *Development of a Solar Powered Solid Absorption Refrigeration System*, Report F30-77, Technical University of Denmark, August 1977.
17. FAVRE, R. and LEIBUNDGUT, H. J., *Ki-Klima-Kälte-Heizung*, 1980, **8**(2), 895–900.
18. KOLBUSCH, P. and SCHÄFER, W., *Biokonversion. Biologisch-Technische Systeme zur Energiegewinnung*, Bundesministerium für Forschung und Technologie, Bonn, 1976.
19. MEYNELL, P.-J., *Methane: Planning a Digester*, Prism Press, Dorset, 1976.
20. BAADER, W. *et al.*, *Biogas in Theorie und Praxis*, Kuratorium für Technik und Bauwesen in der Landwirtschaft (KTBL), e.V., Darmstadt, 1978.
21. BOURNE, M. C., *Journal de l'Institut Canadien de Science et Technologie Alimentaire*, 1973, **6**(4), 81–8.
22. LEPPINGTON, B., *Appropriate Technology*, 1980, **7**(1), 15–18.
23. CHARPENTIER, A. and TUSZYNSKI, W. B., *20th International Dairy Congress*, Congrilait, Paris, 1978.
24. SIEGENTHALER, E. J., *Das Wasserstoffperoxyd-Verfahren als Mittel zur Bereitstellung keimarmer Rohmilch für Käsungsversuche*, ETH-Diss., Juris-Verlag, Zurich, 1965.
25. SIEGENTHALER, E. J. In *Improvement of Livestock Production in Warm Climates* (McDowell, R. E., Ed.), W. H. Freeman and Company, San Francisco, 1972, p. 626.
26. BACHMANN, M., *20th International Dairy Congress, Brief Communications*, Congrilait, Paris, 1978, p. 202.

Chapter 2

COOLING OF HORTICULTURAL PRODUCE WITH HEAT AND MASS TRANSFER BY DIFFUSION

G. van Beek and H. F. Th. Meffert

Sprenger Instituut, Wageningen,
The Netherlands

SUMMARY

The application of physical methods to the prediction of cooling rates of living horticultural produce is a recent development. In living produce physical and biochemical processes, which manifest themselves as mass transfer and heat production within the produce, greatly complicate the application of physical methods to cooling rate calculations. However, by a proper description of these effects in appropriate terms of heat and mass transfer, the laws of thermodynamics can be applied to process calculations for fresh produce.

The variance of biological material is often advocated as a reason for the great differences between calculation and experiment. A closer analysis shows that, in many cases, biological variation is only of secondary importance compared with purely physical phenomena such as the effects of water content, porosity or temperature on the thermal properties involved. On the other hand, certain differences between calculation and experiment may be acceptable in order to simplify a physical model if better accuracy can only be obtained by a much more complicated model and more expensive procedures.

NOMENCLATURE

a	constant
a (m^2/s)	thermal diffusivity

A	constant
A (m^2)	area of surface
b	constant
B	constant
Bi $(\alpha X/\lambda)$	Biot number
c	constant
c (kg/m^3)	water vapour concentration
c_p (J/(kg.K))	specific heat
d (mm or m)	thickness of packaging material
D (m^2/s)	bulk diffusion coefficient
D_{w-a} (m^2/s)	diffusion coefficient of water vapour, in air
E (1/s)	transpiration coefficient
f (h or s)	cooling time from $\theta = 1$ to 0.1
Fo (at/X^2)	Fourier number
H (m/s)	overall coefficient of mass transfer
j	time lag
m (kg)	mass
N	shape factor
Nu $(\alpha X/\lambda)$	Nusselt number
p (Pa)	vapour pressure
P (Pa)	total pressure
Po $(QX^2/\lambda.(T_0 - T_a))$	Pomerantsev number
Pr (v/a)	Prandtl number
q (W/kg)	specific heat generation
q' (W/m^2)	heat flow
Q (W/m^3)	heat generation
r (J/kg)	latent heat of evaporation
R (m)	safe radius
R_d (J/(kg.K))	gas constant of water
Re $\left(\dfrac{vX}{v}\right)$	Reynolds number
s (m)	skin thickness of product
t (h or s)	time
T (°C)	temperature
T_K (K)	temperature
v (m/s)	air velocity
V (m^3)	volume
w (kg/(m^3.s))	evaporation rate
$\dot{w}$ (kg/s)	mass flow of water vapour
x (m)	distance

x_i (kg/kg)	mass fraction of foodstuff component
X (m)	smallest distance from centre to surface
Y (m)	middle distance from centre to surface
Z (m)	longest distance from centre to surface
Z' (h or s)	half cooling time
α (W/(m^2.K))	heat transfer coefficient
β' (m/s)	mass transfer coefficient from surface to air
β	characteristic number
$\Gamma\left(\dfrac{c-c_a}{c_s-c_a}\right)$	dimensionless concentration
ε (m^3/m^3)	porosity or volume fraction
ζ (m^2/m^2)	skin perforation
θ	dimensionless temperature
λ (W/(m.K))	thermal conductivity
μ	permeability
μ'	diffusion resistance factor
ν (m^3/s)	kinematic viscosity
ξ (m^2/m^2)	vent hole area fraction
π (kg/(m.s.Pa))	permeability of packaging material
ρ (kg/m^3)	density
τ	cooling time constant
φ %	relative humidity

Indices

a macro-climate (around box)
c centre
ch centre, due to heat generation
d cooling down period
e end
eff effective
i micro-climate (in box)
o start
s saturated
su surface

1. INTRODUCTION

Keeping perishable produce at low temperatures generally slows down the rate of quality deterioration. Metabolism and microbial growth are

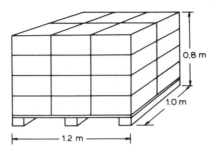

FIG. 1.   Typical dimensions of a pallet load to be cooled.

retarded according to a general rule by a factor of 2–3 per 10 K temperature drop. Even a purely physical process such as transpiration is normally reduced by decreasing the water vapour pressure deficit at lower temperatures.

Cooling of produce from field conditions to storage temperature, is a time consuming process, which needs more time as the available temperature difference between cooling medium and product increases and the exposed surface per volume decreases. The cooling rate is also influenced by the external and internal heat transfer; thus the refrigeration power, needed to perform a cooling process is determined.

In this study, our recent experience with mathematical methods to predict cooling or heating rates and storage conditions of single rectangular units is reviewed, together with their significance for practical application.

The rectangular units are exposed as single bodies to a stream of cooling air. Heat and mass transfer coefficients are assumed to be known (Fig. 1).

Besides heat transfer, mass transfer (especially of water) is considered because of the effect of latent heat transfer on the development of temperature. Attention is also given to heat flow during the cooling period.

These methods range from very simple analytical estimations via graphical means to complicated analytical solutions, which need computer assistance.

Restriction to rectangular units simplifies the mathematical problem to a one-dimensional system; the more dimensional case being a matter of multiplication or addition.

Within these limits one has to deal with thermal and hygric properties of the system, comprising commodity and packaging.

The most simple mathematical expression in terms of heat and mass for living agricultural produce is the respiration equation

$$C_6H_{12}O_6 + 6O_2 \rightarrow 6CO_2 + 6H_2O + 2{\cdot}835 . 10^6 \, J/kmol \, C_6H_{12}O_6 \quad (1)$$

relating the oxidation (combustion) of a nutrient (glucose) to the production of carbon-dioxide and liquid water and the release of heat at 20 °C and atmospheric pressure.

In the practical case, the generated heat can be dissipated by a heat flow, caused by a sensible temperature rise or by evaporation, which can occur as a wet-bulb cooling effect. In any case, the evaporation rate is not related to the water production of the respiration process but to the vapour concentration deficit. However, the more the respiration is restricted the less heat can be dissipated in latent form, which means that the product is forced to heat itself to dispose of the generated heat. A rise in CO_2 and a drop in O_2 concentrations chokes the respiration and consequently heat generation.

2. HEAT TRANSFER THEORY

The physical problem of temperature development in a body under the influence of heat transmission and heat generation can mathematically be expressed by a differential equation

$$\frac{\delta T}{\delta t} = \frac{\lambda}{c_p \rho} \bar{V} T + \frac{Q}{c_p \rho} \tag{2}$$

first formulated by Jean Baptist Fourier in 1822.

For the one-dimensional case—the infinite slab—eqn (2) becomes

$$\frac{\delta T}{\delta t} = a \frac{\delta^2 T}{\delta x^2} + \frac{Q}{c_p \rho} \tag{3}$$

Simple solutions for this differential equation can be found by using the following conditions:

1. The body is homogeneous and isotropic
2. Thermal properties are constant
3. Heat transfer coefficients are constant
4. Initial temperature of the body is uniform
5. Temperature of the cooling medium is constant
6. Heat generation is constant

The exact solution for the temperature field of slab $x = -X < 0 < +X$ with a heat generation Q during cooling has the following form[1]

$$\frac{T - T_a}{T_0 - T_a} = \frac{\text{Po}}{2}\left(1 - \frac{x^2}{X^2} + \frac{2}{\text{Bi}}\right) + \sum_{n=1}^{\infty}\left(1 - \frac{\text{Po}}{\beta_n^2}\right).A_n.\exp(-\beta_n^2 \text{Fo}) \tag{4}$$

where
$$A_n = \frac{2\sin(\beta_n)}{\beta_n + \sin(\beta_n).\cos(\beta_n)} \qquad (5)$$

The solution for the centre temperature (when $x = 0$), responsible for the quality is:

$$\theta = \frac{Po}{2}\left(1 + \frac{2}{Bi}\right) + \sum_{n=1}^{\infty}\left(1 - \frac{Po}{\beta_n^2}\right).A_n.\exp(-\beta_n^2 Fo) \qquad (6)$$

The solution for the mean temperature, responsible for the heat flow, is

$$\theta = \frac{Po}{3}\left(1 + \frac{3}{Bi}\right) + \sum_{n=1}^{\infty}\left(1 - \frac{Po}{\beta_n^2}\right).B_n.\exp(-\beta_n^2 Fo) \qquad (7)$$

where
$$B_n = A_n.\sin(\beta_n)/\beta_n \qquad (8)$$

The necessity of using many roots complicates the calculations, but if Fo > 0.2, only the first root needs to be considered. The centre temperature can be calculated with

$$\theta_c = \frac{Po}{2}\left(1 + \frac{2}{Bi}\right) + \left(1 - \frac{Po}{\beta_1^2}\right).A_1.\exp(-\beta_1^2 Fo) \qquad (9)$$

the mean temperature can be calculated with

$$\theta_m = \frac{Po}{3}\left(1 + \frac{3}{Bi}\right) + \left(1 - \frac{Po}{\beta_1^2}\right).B.\exp(-\beta_1^2 Fo) \qquad (10)$$

and the surface temperature can be calculated with

$$\theta_{su} = \frac{Po}{Bi} + \left(1 - \frac{Po}{\beta_1^2}\right).A_1.\cos\beta_1.\exp(-\beta_1^2 Fo) \qquad (11)$$

where the definitions of the dimensionless numbers are

$$\theta = \frac{T - T_a}{T_o - T_a}: \text{ temperature}$$

$$Po = \frac{QX^2}{\lambda(T_o - T_a)}: \text{ heat generation}$$

$$Fo = \frac{at}{X^2}: \text{ time}$$

$$Bi = \frac{\alpha X}{\lambda}: \text{ ratio between external and internal heat transfer}$$

$$\frac{x}{X}: \text{ position}$$

As calculation from basic formulae is time consuming because of the necessary series expansion, general solutions have been prepared in graphical form. From such graphs, solutions for a particular problem can be produced by simple calculations.

2.1. Cooling Without Heat Generation

Graphical solutions for the temperature development in a slab are normally given for these three special cases:

1. Centre temperature, used to find the slowest cooling in relation to the product quality
2. Mean temperature or enthalpy, necessary to estimate the cooling load
3. Surface temperature, related to cold injury

Of the various forms of presentation, those of Dalgleish and Ede[2] and Pflug and Kopelman[3] deserve special attention.

Dalgleish and Ede presented the complete solution in a set of charts. The principal advantage against earlier presentations, produced between 1920 and 1940 by Gröber, Erk and others, is the possibility of simple extrapolation of the temperature lines to high and low values of Bi, because of their rectilinear asymptotes in the log Fo–log Bi net. Furthermore, the development of the temperature with time can be found on a straight line for Bi = constant (Fig. 2). The temperatures of a multi-dimensional rectangular body are found by multiplication

$$\theta_{xyz} = \theta_x . \theta_y . \theta_z \tag{12}$$

For the initial stage of a cooling process, the resolution is not adequate. Therefore, based on earlier work of Gröber et al.,[13] the solution for a semi-infinite body has been developed following the principle of presentation, adopted by Meffert.[4] This chart (Fig. 3) has the advantage of simple extrapolation and easy estimation of the temperature development in time. But also the temperature distribution at a certain time can easily be found along a straight line with the slope -2 in the log–log network. It is possible to use this chart without loss of accuracy until the temperature rise from one side has reached the opposite surface according to the 'rule of the inversed tail'.

For the calculation of heat flow from the surface of a semi-infinite body

$$q' = \alpha . (T_s - T_a) \tag{13}$$

the surface temperature has to be determined. The dimensionless surface

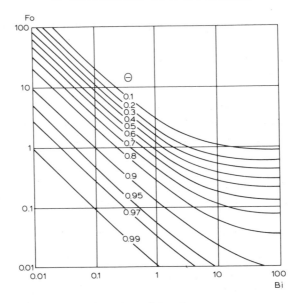

FIG. 2.   Chart of Dalgleish and Ede[2] for the centre temperature of a slab.

temperature is given in Fig. 4 by Grigull.[14] For practical calculations, this curve can be replaced by the following equation

$$\theta_{su} = A \cdot \exp\left(B \cdot \frac{\alpha}{x} \sqrt{a \cdot t}\right) \qquad (14)$$

where the constants have the values $A = 0.982$, $B = -0.919$ if $\theta_s > 0.3$; $A = 0.409$, $B = -0.218$ if $\theta_s < 0.3$.

Pflug and Kopelman[3] elaborated another type of graphical solution by concentrating on the exponential part of the general solutions. They proposed to define the cooling process by the rate of exponential cooling and a lag factor, which fixes the course of temperature in time. Pflug and Kopelman, therefore, can produce with one presentation of Fo versus Bi and a presentation of the intercepts the basic information on centre, mean and surface temperature.

All these types of graph allow easy extrapolation to higher and lower Bi numbers.

The first root approach without heat generation, yields for the centre temperature:

$$\theta_c = A_1 \cdot \exp(-\beta_1^2 Fo) = j_c \cdot \exp(-\beta_1^2 Fo) \qquad (15)$$

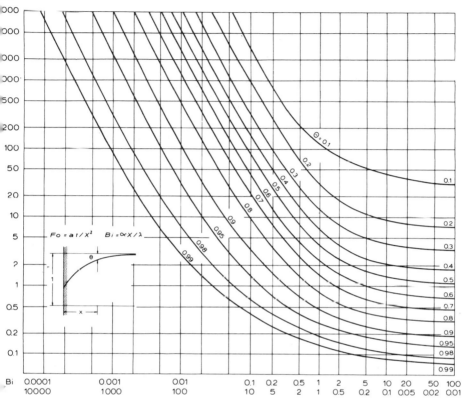

FIG. 3. Chart of Meffert for the temperature in a semi-infinite body.

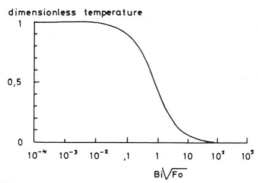

FIG. 4. Dimensionless temperature of the surface of a slab.

for the mean temperature:

$$\theta_m = B_1 . \exp(-\beta_1^2 Fo) = j_m . \exp(-\beta_1^2 Fo) \tag{16}$$

and for the surface temperature:

$$\theta_{su} = A_1 . \cos\beta_1 . \exp(-\beta_1^2 Fo) = j_s . \exp(-\beta_1^2 Fo) \tag{17}$$

The common term represents the cooling rate of the exponential part of the process.

In a semilog plot, eqns (15), (16) and (17) are represented by parallel lines (see Fig. 11, Section 2.2.4)

$$\log\theta = -\frac{t}{f} + \log j \tag{18}$$

with the slope

$$\frac{1}{f} = \frac{\beta_1^2 . a}{2 \cdot 303 X^2} \tag{19}$$

and the intercept value j at $t = 0$.

The dimensionless presentation

$$\frac{fa}{X^2} = \frac{2 \cdot 303}{\beta_1^2} \tag{20}$$

and j are only functions of Bi and are given in Figs. 5 and 6, respectively. The $1/10$ exponential cooling time is equal to f.

The course of $2 \cdot 303/\beta_1^2$ can be traced in the Dalgleish–Ede chart for $\theta_c = 0 \cdot 1$.

For three-dimensional problems the solution is found by

$$\frac{1}{f_{xyz}} = \frac{1}{f_x} + \frac{1}{f_y} + \frac{1}{f_z} \tag{21}$$

and

$$j_{xyz} = j_x . j_y . j_z \tag{22}$$

according to the contribution of the heat transfer in the three dimensions. The cooling time becomes:

$$t = f_{xyz} . \log\left(j_{xyz}\frac{1}{\theta}\right) \tag{23}$$

Again, the initial phase of the cooling process is not covered by this approach. A reasonable accuracy is reached beyond $\theta = 0 \cdot 6$.

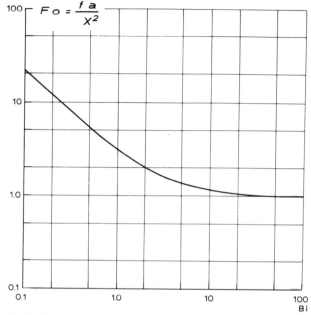

FIG. 5. Relation between dimensionless heat transfer coefficient and time of a slab.

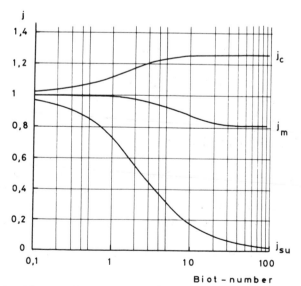

FIG. 6. Time lag factors of centre, mean and surface of a slab.

TABLE 1

	f	Z'	$\dfrac{1}{C_r}$	τ
f	—	3·32	2·303	2·303
Z'	0·302	—	0·693	0·639
$\dfrac{1}{C_r}$	$\dfrac{1}{2·303}$	$\dfrac{1}{0·693}$	—	1
τ	0·4342	1·443	1	—

$f = 1/10$ cooling time (h)
$Z' = 1/2$ cooling time (h)
$\dfrac{1}{C_r} =$ cooling rate $\left(\dfrac{1}{h}\right)$
$\tau =$ cooling time constant (h)

The advantage of this approach, however, goes further than the simple graphical presentation of the basic solution. By accepting the simplification[5]

$$\frac{fa}{X^2} \simeq 1 + \frac{2}{Bi} \tag{24}$$

instead of

$$\frac{fa}{X} = \frac{\ln 10}{2}\left(1 + \frac{2}{Bi}\right) \tag{25}$$

the 1/10 cooling time can be estimated by a very easy calculation. This simplification allows also a convenient approach of the cooling process with heat generation, showing that this can be covered by the same principles.

The 1/10 cooling time is related to other expressions, used to describe the cooling rate by the relations shown in Table 1.

2.2. Cooling With Heat Generation

2.2.1. Initial Temperature Rise and Residual Temperature Difference

The cooling process of heat generating bodies is characterised by an initial rise of the temperature, which can result in a thermal explosion, if certain criteria for the cooling process are not satisfied. After cooling down there remains a residual temperature difference between body and cooling medium.

The initial temperature rise follows from the basic differential equation

$$\frac{\delta T}{\delta t} = a\frac{\delta^2 T}{\delta x^2} + \frac{Q}{c_p \rho} \tag{26}$$

for we find, if $t = 0$ and $x = 0$, $\delta^2 T/\delta x^2 = 0$, i.e. uniform temperature throughout the body at the beginning of the process

$$\frac{dT}{dt} = \frac{Q}{c_p \rho} \tag{27}$$

This shows that the initial temperature rise is a product property containing only specific data of the product called the adiabatic heating rate.

At the end of the, cooling process we can introduce the residual temperature difference. After a long time—large Fo numbers—eqns (9), (10) and (11) become

centre temperature

$$\theta_{cc} = \frac{Po}{2}\left(1 + \frac{2}{Bi}\right) \tag{28}$$

mean temperature

$$\theta_{me} = \frac{Po}{3}\left(1 + \frac{3}{Bi}\right) \tag{29}$$

surface temperature

$$\theta_{sue} = \frac{Po}{Bi} \tag{30}$$

from which, by inserting θ and Po, expressions for the temperature difference are derived

centre temperature

$$(T_e - T_a)_c = \frac{QX^2}{2\lambda}\left(1 + \frac{2}{Bi}\right) \tag{31}$$

mean temperature

$$(T_e - T_a)_m = \frac{QX^2}{\lambda}\frac{1}{3}\left(1 + \frac{3}{Bi}\right) \tag{32}$$

surface temperature

$$(T_e - T_a)_{su} = \frac{QX^2}{\lambda}\frac{1}{Bi} \tag{33}$$

By the following simplification

$$\frac{fa}{X^2} = 1 + \frac{2}{Bi} \tag{34}$$

the residual centre temperature difference becomes

$$(T_e - T_a)_c = \frac{Q}{2\lambda} \cdot fa \tag{35}$$

This equation relates residual centre temperature to cooling time.[5] This relation has been anticipated by Sainsbury[6] and again described by Lindau and Matthisen.[7]

2.2.2. First Root Solution for a Slab

The first root solution is based on the exponential cooling curve. At the beginning the temperature fields are not developed. We can introduce a time lag for heat generation by rearrangement of eqn (9)

$$\theta_c - \frac{Po}{2}\left(1 + \frac{2}{Bi}\right) = \left(1 - \frac{Po}{\beta_1^2}\right).A.\exp(-\beta_1^2 Fo) \tag{36}$$

showing that

$$\theta_c - \theta_{ce} = \left(1 - \frac{Po}{\beta_1^2}\right).A_1.\exp(-\beta_1^2 Fo) \tag{37}$$

Dividing by $1 - \theta_{ce}$, eqn (37) becomes

$$\frac{\theta_c - \theta_{ce}}{1 - \theta_{ce}} = \frac{1 - \dfrac{Po}{\beta_1^2}}{1 - \dfrac{Po}{2}\left(1 + \dfrac{2}{Bi}\right)}.A_1.\exp(-\beta_1^2 Fo) \tag{38}$$

or

$$\frac{T - T_e}{T_o - T_e} = j_{ch} \cdot j_c \cdot \exp(-\beta_1^2 Fo) \tag{39}$$

which formally corresponds with eqn (15) if

$$j_{ch} = \frac{1 - \dfrac{Po}{\beta_1^2}}{1 - \dfrac{Po}{2}\left(1 + \dfrac{2}{Bi}\right)} \tag{40}$$

This indicates that the influence of heat generation on cooling time can be expressed as another lag factor in a first root solution. The values for β_1 can be taken from Fig. 7 or convenient tables.[1]

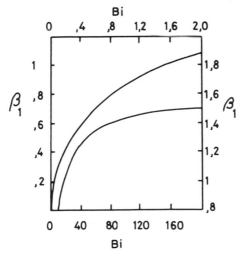

FIG. 7. The value of the characteristic number β_1.

Using the relation in eqn (20) and the simplification in eqn (25), it can be shown that

$$\frac{1}{\beta_1^2} = \frac{1}{2 \cdot 303} \frac{f.a}{X^2} \qquad (41)$$

in which case the lag factor due to heat generation becomes

$$j_{ch} = \frac{1 - \dfrac{Po}{2 \cdot 303} \dfrac{f.a}{X^2}}{1 - \dfrac{Po}{2}\left(1 + \dfrac{2}{Bi}\right)} \qquad (42)$$

From this formulation eqn (40) acts as a criterion for the possibility of cooling down. It can be derived that cooling is only possible if

$$\frac{Po}{2}\left(1 + \frac{2}{Bi}\right) \le 1 \qquad (43)$$

which holds when the time lag due to heat generation becomes excessive. For large Bi numbers, the criterion for cooling becomes simply $Po < 2$.

2.2.3. 'Safe Radius'
From eqn (27) it is possible to define a criterion for the size of a body in which a certain temperature difference is not exceeded: the 'safe radius'.

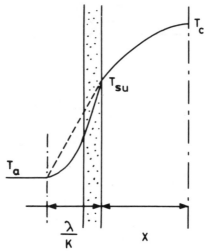

FIG. 8.   Temperature profile at the surface.

The temperature difference between the centre of a heat-generating body and the surrounding medium as presented in eqn (27) contains two terms (see Fig. 8)

in the product:

$$T_c - T_{su} = \frac{QX^2}{2\lambda} \tag{44}$$

and between surface and medium:

$$T_{su} - T_a = \frac{QX}{\alpha} \tag{45}$$

The 'safe radius' for keeping the temperature difference in a product unit, lower than a safe value, is given by

$$R_{\Delta T} = \sqrt{\frac{2\lambda}{Q} \cdot \Delta T} \tag{46}$$

This relation is also valid for heat generation increasing with temperature. In that case heat generation has to be taken at the maximum level $Q_{T+\Delta T}$. Figure 9 shows the graphical solution of this problem.

A 'standard safe radius' can be formulated for $\Delta T = 1\,°C$, taking the product properties at a temperature of $T + 1$:

$$R = \sqrt{\frac{2\lambda}{Q_{(T+1)}}} \tag{47}$$

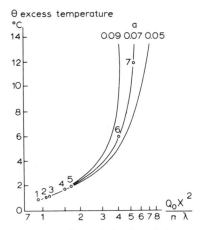

FIG. 9. Excess temperature of a unit load with heat-generating produce.

2.2.4. First Root Solution of Rectangular Bodies

The multi-dimensional cooling down problem with the first root solution for rectangular bodies uses a shape factor N as given in Fig. 10. It is not possible to use the multiplication $\theta_{xyz} = \theta_x . \theta_y . \theta_z$ with heat-generating products at high Fo numbers with only six or less characteristic values of β.

With reference to the smallest dimension, eqn 9 becomes

$$\theta_c = \frac{Po}{N}\left(1 + \frac{2}{Bi}\right) + \left(1 - \frac{Po}{\beta_1^2}\right).A_1.\exp(-\beta_1^2 Fo) \tag{48}$$

The residual temperature corresponds to

$$\theta_{ce} = \frac{Po}{N}\left(1 + \frac{2}{Bi}\right) \tag{49}$$

or

$$\theta_{ce} = \frac{QX^2}{N\lambda}\left(1 + \frac{2}{Bi}\right) \tag{50}$$

The time lag factor due to heat generation is

$$j_{ch} = \frac{1 - \dfrac{Po}{\beta_1^2}}{1 - \dfrac{Po}{N}\left(1 + \dfrac{2}{Bi}\right)} \tag{51}$$

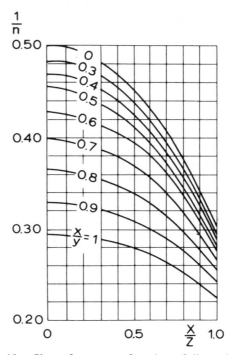

FIG. 10.   Shape factor as a function of dimensions.

where

$$\frac{1}{\beta_1^2} = \frac{1}{\beta_1 X^2} + \left(\frac{X}{Y}\right)^2 \frac{1}{\beta_1 X^2} + \left(\frac{X}{Y}\right)^2 \frac{1}{\beta_1 X^2} \tag{52}$$

Using the simplification of eqns (25) and (41), the time lag factor due to heat generation becomes

$$j_{ch} = \frac{1 - \dfrac{Po_x}{2 \cdot 303} \dfrac{a}{X^2} \cdot f_{xyz}}{1 - \dfrac{Po_x}{N}\left(1 + \dfrac{2}{Bi_x}\right)} \tag{53}$$

where f_{xyz} is calculated by eqn (21). Figure 11 shows a presentation of the first root solution with heat generation.

From the simplified expression for the dimensionless 1/10 cooling time

$$\frac{fa}{X} = 1 + \frac{2}{Bi} \tag{54}$$

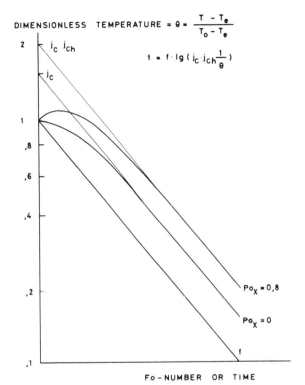

FIG. 11. Example of cooling curve according to Pflug and Kopelman.[3]

TABLE 2
RELATION BETWEEN 1/10 COOLING TIME AND RESIDUAL
TEMPERATURE FOR RECTANGULAR BODIES

Geometry	$f\dfrac{a}{X^2}$	$\theta_{ce}\dfrac{a}{X^2}\dfrac{c_p}{Q}$	$\dfrac{\theta_{ce}c_p}{fQ}$
Slab	$1+\dfrac{2}{Bi}$	$1/2\left(1+\dfrac{2}{Bi}\right)$	$1/2$
Square rod	$1/2\left(1+\dfrac{2}{Bi}\right)$	$1/3\cdot4\left(1+\dfrac{2}{Bi}\right)$	$1/1\cdot7$
Cube	$1/3\left(1+\dfrac{2}{Bi}\right)$	$1/4\cdot5\left(1+\dfrac{2}{Bi}\right)$	$1/1\cdot5$

and the eqn (34) for the residual temperature

$$\theta_{ce} = \frac{QX^2}{N\lambda}\left(1 + \frac{2}{Bi}\right) \tag{55}$$

a simple relation can be formed between cooling rate and residual temperature:

$$\theta_{ce} = \frac{Q}{N}\frac{a}{\lambda}f = \frac{Qf}{Nc_p} \tag{56}$$

As N and f are dependent on the geometry of the body in different ways, f according to eqn (21) and N (presented by Tchumak et al.[8]), the relation θ_{ce}/f can only be given for certain geometries (see Table 2).

3. MASS TRANSFER THEORY

Transfer of water vapour is an important contribution to the transfer of heat from horticultural produce.

During the cooling process, evaporation hastens the cooling rate at the expense of mass loss. The keeping condition for the product (micro-climate) is strongly influenced by mass transfer effects. Cooling of one tomato is faster than heating the same tomato, as shown in Fig. 12. Fockens and Meffert[9] proposed a three-fold physical model for the evaporation from horticultural produce (type IV in Fig. 13).

Mass transfer of a product during a cooling process, depends much on the properties of its skin. Produce with a wet surface (type I) shows a relative mass loss of

$$\frac{\Delta m}{m} = \frac{\Delta T_d \cdot cp}{R_D T_K \left[\frac{\alpha \int_0^{\Delta td}(T_s - T_a)\,dt}{\beta' \int_0^{\Delta td}(p_s - p_a)\,dt} + r\right]} \tag{57}$$

The numerical value of the quotient of integrals in the demominator of eqn (57), is almost constant in the temperature range where vapour pressure can be taken as proportional to the temperature. Equation (57) shows that:

1. Mass loss during cooling does not depend on the cooling time.
2. Mass loss during cooling does not depend on air velocity as transfer coefficients α (heat) and β' (mass) range within the same rate.
3. Mass loss during cooling is proportional to temperature drop.

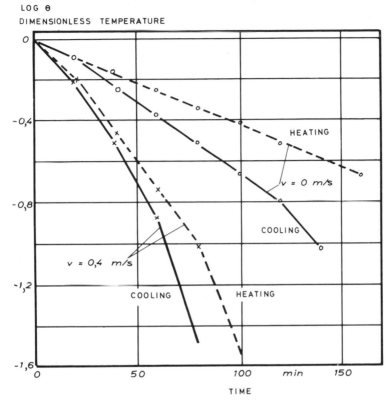

FIG. 12. Comparison of cooling and heating of one tomato.

The relative mass loss during the cooling of a product with a skin of type II can be given as

$$\frac{\Delta m}{m} = \frac{\Delta T_d \cdot c_p}{\dfrac{\alpha R_0 T_K \mu s}{\zeta_2} \cdot \dfrac{\int_0^{\Delta td} (T_s - T_a)\,dt}{\int_0^{\Delta td} (p_s - p_a)\,dt} + r} \tag{58}$$

Again taking the quotient of integrals as constant, the relative mass loss depends on the heat transfer coefficient α. This means that by increasing heat transfer the mass loss becomes smaller. For produce with a skin of high diffusion resistance, mass loss decreases with cooling time and with higher air velocities. Mass loss is proportional to temperature drop.

Cooling the contents of a non-ventilated box from the outside, results normally in condensation conditions against the inner lining of the package

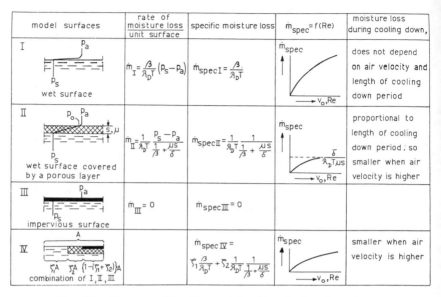

FIG. 13. Four possible skin models of horticultural produce.

(see the example in Section 5.3 with potatoes). An absorptive package material, a low diffusion resistance of the walls, or slight ventilation can overcome this kind of problem.

It still remains, that a long cooling time of large unit loads results in considerable mass losses. Reducing cooling time by subdivision of units or ventilation reduces mass loss during the cooling period. Mathematical modelling of the heat and mass transfer of horticultural produce during the cooling process, normally leads to complicated computer models. These are restricted in their predictional power.

For the prediction of keeping conditions in terms of water vapour concentration, Cowell[10] has developed an analytical approach. The local concentration difference in a slab of a homogeneous granular solid covered with a membrane of permeability μ, can be given as

$$\Gamma = \frac{c - c_a}{c_s - c_a} = 1 - \frac{\dfrac{\mu}{D}\cosh\left(x\sqrt{\dfrac{E}{D}}\right)}{\dfrac{\mu}{D}\cosh\left(X\sqrt{\dfrac{E}{D}}\right) + \sqrt{\dfrac{E}{D}}\sinh\left(X\sqrt{\dfrac{E}{D}}\right)} \tag{59}$$

Figure 14 shows calculated results of currants in boxes, indicating the influence of box dimensions and vent hole area.

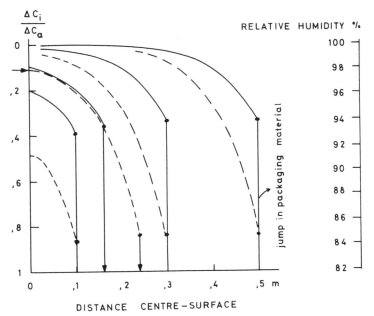

FIG. 14.   Water vapour concentration in a slab. Vent hole area: ———, 10%; – – –,
100%.

The evaporation rate of a slab is represented by

$$w = E.(c_s - \bar{c})$$

(60)

where $\bar{c}$ is the mean water vapour concentration (kg/m³).
   The mean water vapour concentration deficit is

$$c_s - \bar{c} = \frac{c_s - c_a}{X\sqrt{\dfrac{E}{D}}\operatorname{cotanh}\left(X\sqrt{\dfrac{E}{D}}\right) + X\dfrac{E}{H}}$$

(61)

thus, the substitution of eqn (61) in eqn (60) yields the evaporation rate. As a
first practical approach to the calculation of the concentration difference,
multiplication in the relevant dimensions seems to be justified. The
concentration in a rectangular body is analogous to the solution of the
thermal problem

$$\Gamma_{XYZ} = \Gamma_X.\Gamma_Y.\Gamma_Z$$

4. THERMAL AND HYGRIC PROPERTIES

4.1. Prediction of Thermal Properties

Measured thermo-physical properties of horticultural produce in bulk are rarely reported in the literature, but the composition is nearly always known. The mass fractions of water, carbohydrate, protein and fat can be taken from several sources. Thermo-physical properties of agricultural produce can be calculated on this basis. Specific heat is calculated with eqn (62) using the properties of foodstuff components given in Table 3.

$$c_{\text{product}} = \sum x_i \cdot c_{p_i} \tag{62}$$

The influence of small deviations in the (calculated) specific heat on cooling time is negligible in contrast to the next term.

TABLE 3
PHYSICAL PROPERTIES OF FOODSTUFF COMPONENTS AT $0\,°C$

Component	ρ (kg/m^3)	c_p $(J/(kg.K))$	λ $(W/(m.K))$
Water	1·000	4·217	0·569
Carbohydrate	1·550	1·220	0·28
Protein	1·380	1·900	0·20
Fat	0·930	1·900	0·18
Minerals	2·400	0·840	—
Air	1·29	1·005	0·024 5

Thermal conductivity is calculated according to the parallel heat flow theory. This theory applies to the volume fraction of the components of foodstuffs. The volume fraction of solid components can be found as follows

$$\varepsilon_i = \frac{\rho_{\text{bulk}}}{\rho_i} x_i \tag{63}$$

ε_i = volume fraction of component (m^3/m^3)
ρ_{bulk} = density of foodstuff in bulk (kg/m^3)
ρ_i = density of component (kg/m^3)
x_i = mass fraction of component (kg/kg)

The volume fraction or porosity of gas components, in general air, is the deficit value

$$\varepsilon_{\text{total}} = 1 - \sum \varepsilon_i \tag{64}$$

The total volume fraction of air in the systems consists of the porosity of the single product and the porosity of the bulk. The porosity of a single product is

$$\varepsilon_{product} = 1 - \frac{\rho_{measured\ product}}{\rho_{calculated}} \tag{65}$$

where the denominator is the density of a product without air spaces. Only the mass fraction of the components is necessary to find this theoretical product density

$$\rho_{calculated} = \frac{1}{\sum \frac{x_i}{\rho_i}} \tag{66}$$

The porosity of the bulk is

$$\varepsilon_{bulk} = 1 - \frac{\rho_{bulk}}{\rho_{product}} \tag{67}$$

Unfortunately, it is not possible to find the total porosity by adding the bulk and product porosity. To find the total porosity the next equation

$$\varepsilon_{total} = \varepsilon_{product} \cdot \frac{\rho_{bulk}}{\rho_{product}} + \varepsilon_{bulk} \tag{68}$$

gives the same result as eqn (64).

Another equation for the total porosity is

$$\varepsilon_{total} = 1 - \frac{\rho_{bulk}}{\rho_{calculated}} \tag{69}$$

Table 4 shows the porosity of the product and the bulk density and

TABLE 4
DENSITIES AND POROSITY OF HORTICULTURAL PRODUCE

Name of product	Density		Porosity		
	product (kg/m^3)	bulk (kg/m^3)	product $(\%)$	bulk $(\%)$	total $(\%)$
Apple	779	500	27	35	52
Pear	990	600	6	39	43
Flower bulbs	880	600	16	32	43
Cut flowers	780	200	23	74	80
Sweet pepper	480	260	53	46	75
Berries (red)	1 050	580	0	54	45
Potato	1 075	590	0	45	45

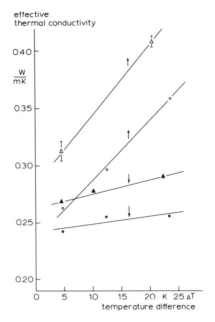

FIG. 15.   Effect of natural convection on thermal conductivity of eggs, ($\triangle$, heat flow up; $\blacktriangle$, heat flow down) and artificial lemons ($\bigcirc$, heat flow up; $\bullet$, heat flow down) in bulk.

measured densities of some agricultural products. The thermal conductivity follows from

$$\lambda = \sum \varepsilon_i \lambda_i \tag{70}$$

The gas volume fraction has an important influence on the actual thermal conductivity. This calculation method cannot take into account the heat flow direction. Figure 15 shows that thermal conductivity of bulk packed eggs is always higher when the heat flow has been directed upwards, owing to the influence of natural convection.

With this method the thermo-physical properties of anisotropic materials cannot be calculated properly.

Thermal diffusivity is an important property in cooling problems

$$a = \frac{\lambda}{c_p \cdot \rho} \tag{71}$$

It is necessary to determine the thermo-physical properties as accurately as

TABLE 5
THERMO-PHYSICAL PROPERTIES OF PRODUCTS

Product	λ $(W/(m.K))$	ρ (kg/m^3)	c_p $(J/(kg.K)$	a $(m^2/s.10^{-7})$
Eggs	0·310	600	3 654	1·41
Flower bulbs	0·335	600	3 804	1·47
Pear	0·333	600	3 771	1·47
Cut flowers	0·136	200	4 081	1·66
Sweet pepper	0·174	275	3 984	1·59
Berries	0·323	580	3 778	1·47
Apple	0·286	500	3 827	1·49
Potato	0·311	590	3 571	1·47

possible; about $\pm 5\%$ to $\pm 15\%$ to achieve a reasonable accurate value for the thermal diffusivity.

In conclusion, Table 5 shows the calculated thermo-physical properties of products to be used in the examples of Section 4.

4.2. K-value of Packaging and Cooling Time

The influence of the packaging on cooling time is expressed in the Biot number. The heat transfer coefficient can be given as

$$\alpha = \frac{1}{\dfrac{1}{\alpha_{convection}} + \left(\dfrac{d}{\lambda}\right)_{packaging}} \tag{72}$$

$\alpha_{convection}$ can be found by

$$Nu = c\,Re^a.\,Pr^b \tag{73}$$

The value of the constants a, b and c are given in the literature. For a smooth plane with air flow parallel to the surface

$$\alpha_{convection} = 5\cdot 8 + 3\cdot 9.v \qquad (v < 5\,m/s) \tag{74}$$

During cooling of horticultural produce, air velocities between 0·4 and 2 m/s are normally used. The thickness of the packaging material ranges from 0 to 0·03 m. The thermal conductivity of some packaging materials are given in Table 6 (where λ/d is the K-value of the packaging).

In Fig. 16 the heat transfer coefficient is given as a function of the packaging material thickness at an air velocity v of 0·5 m/s.

Metal has no significant influence on heat transfer, the corrugated carton

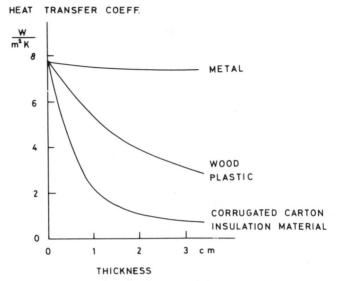

FIG. 16. Influence of packaging material on heat transfer coefficient.

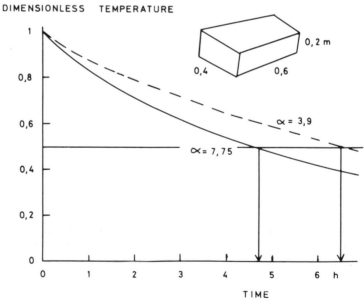

FIG. 17. A lower heat transfer coefficient decreases cooling rate.

TABLE 6
THERMAL CONDUCTIVITY OF PACKAGING
MATERIALS

Material	λ $(W/(m.K))$
Wood	0·16
Paper	0·13
Polyethylene (LD)	1·35
Polyethylene (HD)	0·50
Plastics	0·15
Insulation materials	0·030
Corrugated carton	0·044

and insulation material determine it. Wood of 2-cm thickness halves α while corrugated carton and insulation material reduces α by more than 50%. In theory the heat transfer coefficient is part of the Biot number. When Bi > 10, a change of Bi has no significant influence on the cooling time.

Figure 17 shows the calculated influence of packaging material on cooling. Reducing α to 3·9 gives a half cooling time of 6·4 h compared to 4·7 h with a heat transfer coefficient of 7·75. In this example the Bi number is less than 10.

The theory of cooling unit loads without internal air circulation, does not give an answer to the question of the relationship between vent holes and cooling time. The only way to introduce vent holes is calculation of the heat transfer coefficient with the vent hole area fraction

$$\alpha = \frac{1}{\dfrac{1}{\alpha_{\text{convection}}} + \dfrac{1}{\xi\left(\dfrac{\lambda}{d}\right)_{\text{air}} + (1-\xi).\left(\dfrac{\lambda}{d}\right)_{\text{mat}}}} \tag{75}$$

This equation represents the heat transfer for a vent hole arrangement, shown in Fig. 18.

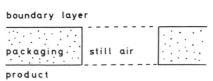

FIG. 18. Physical model of a vent hole.

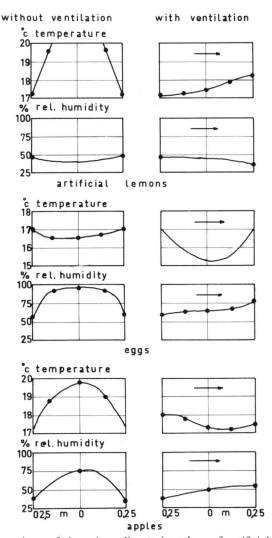

FIG. 19. Comparison of the micro-climate in a box of artificial lemons (heat generation only), of eggs (evaporation only) and apples (heat generation and evaporation). A small air velocity of 3 mm/s in the box shifts the profile to the right.

TABLE 7
INFLUENCE OF VENT HOLES IN A CORRUGATED CARTON AND
METAL ON THE HEAT TRANSFER COEFFICIENT

	Corrugated carton		Metal	
$d\,(mm)$	3	1	3	1
$\lambda\,(W/(m.K))$	0·044	0·044	50	50
Vent hole area (%)				
30	4·92	6·62	8·00	8·00
10	5·10	6·72	8·00	8·00
0	5·18	6·77	8·00	8·00

$\alpha_{\text{convection}} = 8\ W/(m^2.K)$.

Introducing realistic values in eqn (75) shows that vent holes do not determine the overall heat transfer coefficient. Table 7 illustrates this point. Nevertheless the cooling time of unit loads is determined by the percentage of vent holes. Small pressure differences around a unit load introduce low air velocities in the unit load. This convection heat transfer is enough to hasten the cooling of unit loads. Figure 19 illustrates the effect of a small air velocity (3 mm/s) through the unit load on temperature and relative humidity profiles. It shows that the maximum is lowered and shifted into the direction of the air flow.

Another example is given in Fig. 20. It indicates that a wrapped unit load has twice the 90 % cooling time of an unwrapped unit load. This result was found in four experiments using pears, flower bulbs, artificial lemons and tomatoes. The relation between percentage of vent holes in the packaging and half cooling time is measured in experiments using two types of stacking arrangements. Figure 21 shows the results of five packagings in a pile and Fig. 22 shows the results of pallet loads with dimensions of $1 \times 1·2 \times 1·5$ m. The minimum effect of the vent holes in the packaging is given by the straight line.

4.3. Heat Generation

The heat generation of the respiration process is a well documented thermal property of living produce. However, the large spread of experimental data, and details lost by the long chain of references, do not contribute to the accuracy of predictive calculation.

The heat of respiration is usually calculated from measured carbon dioxide production. The value of this method is limited. The respiration equation (Section 1) gives the mass and energy balances for the combustion

product flowerbulbs in boxes

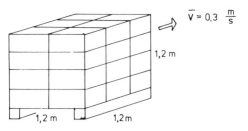

demensionless temperature of the centre of the load

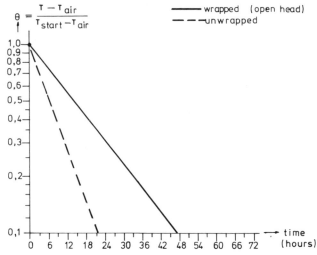

FIG. 20. Wrapping of a unit load doubles the half cooling time.

of solid glucose. The respiration coefficient O_2/CO_2 is generally not equal to unity because in horticultural produce glucose is not the only source of energy. Furthermore, the energy content of dissolved glucose is different. Because of the variation of the conversion factor heat/CO_2 with temperature, results from respiration measurements have to be looked upon with great care. The physiological age of the sample at the moment of measurement may also be of influence. Direct measurement of the heat generation rate is preferable. An adiabatic calorimeter, allowing for measurement as a function of temperature for different conditions of humidity and gas composition was described by Rudolphij et al.[11] Figure 23 shows the results of measurements with the adiabatic calorimeter. The

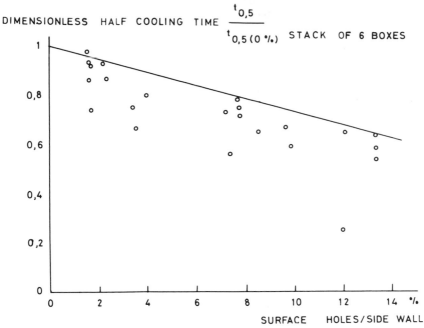

FIG. 21. Effect of vent holes on half cooling time on a stack of six boxes.

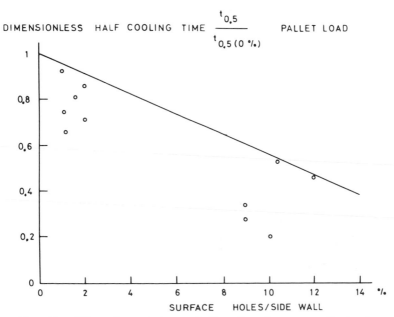

FIG. 22. Effect of vent holes on half cooling time on a pallet load.

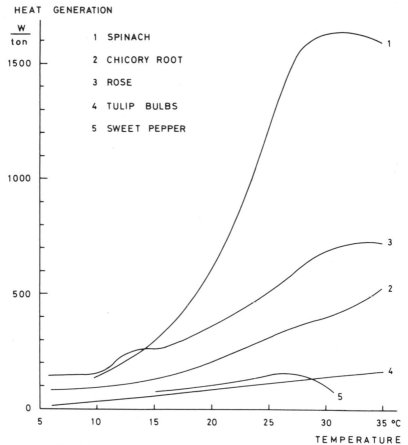

FIG. 23.   Heat generation of some horticultural produce.

temperature range above 30 °C seems to be detrimental to leafy products. In previous measurements, flower bulbs showed a decreasing heat generation rate around 50 °C.

For the direct measurement, the effect of evaporation has to be considered carefully. The effective heat generation, which is lower than the heat of respiration, can be found from the mass loss during the period of observation. This mass loss consists of carbon loss by oxidation and water loss by evaporation.

The effective heat generation rate is given by

$$q_{\text{eff}} = 1 \cdot 063q - 2 \cdot 48 . 10^{-6} \left(\frac{\Delta m}{m \, \Delta t} \right) \tag{76}$$

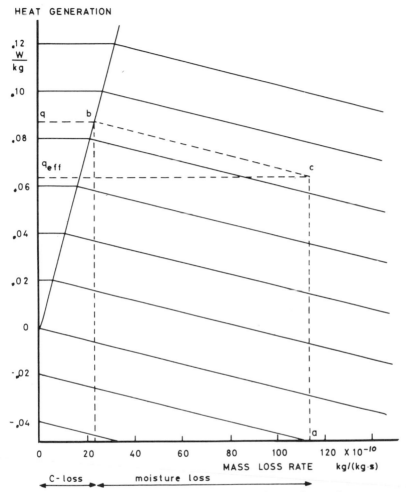

FIG. 24. Nomogram relating heat generation and mass loss.

This relation is shown in Fig. 24 in which the falling parallel lines depict this equation. The specific heat generation q is shown on the figure, as is the horizontal distance to b, which is the carbon loss. From the mass loss at a and the intercept c, the effective heat generation q_{eff} is found.

4.4. Hygric Properties

The transpiration hastens the cooling of horticultural produce. Figure 12 gives the comparison of the cooling and heating curves of one tomato. It is

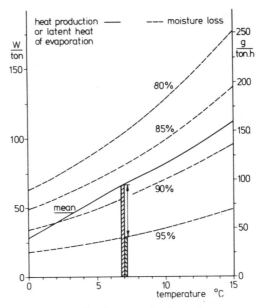

FIG. 25. The effective heat production of sweet pepper.

clear that the slope of the cooling curve is steeper than that of the heating curve, indicating the presence of a latent heat effect by transpiration. Transpiration reduces the heat production of the living produce. The effective heat production is

$$Q_{\text{eff}} = Q - r.w \tag{77}$$

where Q_{eff} = effective heat production (W/m³)
   Q = heat production (W/m³)
   r = latent heat of evaporation (J/kg)
   w = evaporation rate (kg/(m³.s))

Figure 25 shows a presentation of the heat generation of sweet peppers as a function of temperature and moisture loss. Sweet pepper stored at a temperature of 7 °C and a relative humidity of 90 % has an effective heat generation Q_{eff} of $76 - 33 = 43$ W/ton.

The transpiration rate of horticultural produce is proportional to water vapour concentration deficit, the driving force of transpiration

$$w = E.(c_s - c) \tag{78}$$

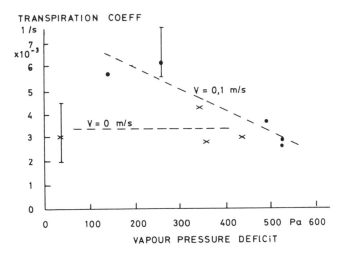

FIG. 26. The transpiration coefficient of pear is not a constant.

where E = transpiration coefficient for bulk (1/s)

c_s = concentration of water vapour in the product (kg/m^3)

c = concentration of water vapour around the product (kg/m^3)

The transpiration coefficient for bulk is not constant but depends on air velocity, concentration deficit and, of course, the properties of the skin. Figure 26 gives an example of the transpiration coefficient for the pear. At low air velocities, E is not a function of the concentration deficit.

The water vapour concentration in the product is near the saturated concentration c_s.

The equilibrium relative humidity is 98·5–100 %. In general it is possible to estimate the equilibrium relative humidity from the initial freezing point of the foodstuff, every °C below 0 °C gives a lowering of nearly 1 %.

The concentration of water vapour around the product in the so-called micro-climate is a function of

1. The transpiration coefficient of the product
2. The packaging
3. The stacking pattern of the packagings
4. The macro-climate

The air around the unit load is called the macro-climate (see Fig. 27). The theoretical approach to predict the concentration profile in a unit load is given by Cowell.[10]

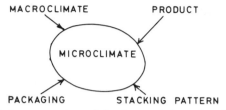

FIG. 27.   Graphical presentation of the factors influencing the micro-climate.

The bulk transpiration coefficients of the products used in the examples are given in Table 8. The dimension 1/s of this coefficient seems meaningless, but looking at eqn (79), it is clear that it is correct

$$\frac{kg}{m.\dfrac{kg}{m}.s} = \frac{1}{s} \tag{79}$$

Coming to the relation between macro- and micro-climate, the diffusion coefficient of water vapour in air has to be estimated. The Schirmer equation gives this coefficient as a function of pressure and temperature

$$D_{w-a} = 23 \cdot 10^{-6} \cdot \left(\frac{10^5}{P}\right) \cdot \left(\frac{T_K}{273}\right) 1\cdot81 \tag{80}$$

For a bulk of products the diffusion coefficient depends on the bulk porosity

$$D = D_{w-a} \cdot (1 - \sqrt{(1 - \varepsilon)}) \tag{81}$$

TABLE 8

TRANSPIRATION COEFFICIENT OF SOME PRODUCTS
AND THEIR BULK DIFFUSION COEFFICIENTS

Product	E $(1/s.10^{-3})$	D $(m^2/s.10^{-6})$
Egg	1·10	—
Flower bulbs	1·14	4·03
Pear	3·26	5·04
Cut flowers[a]	32·2	11·3
Sweet pepper	5·42	6·10
Berries	10·0	7·40
Apple	2·88	4·46
Potato	1·54	5·94

[a] This data is given for tulip species.

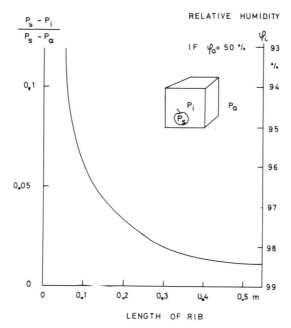

FIG. 28. Water vapour pressure and relative humidity in packages of cartons without vent holes.

This diffusion coefficient can be too low in many practical cooling experiments, because convection (natural or forced) increases the coefficient by a factor of 2–4.

The bulk porosity is given in Table 4. The calculated bulk diffusion coefficients are given in Table 8.

Vent holes in the packaging chiefly determine the relation between packaging and micro-climate. The mass transfer through packaging without vent holes is very small, resulting in nearly saturated water vapour concentration in this type of packaging. The dimensionless water vapour concentration in a package without vent holes, assuming no concentration gradients in the unit load, is

$$\frac{c_s - c_i}{c_i - c_u} = \frac{\pi A}{dV} = \frac{7 \cdot 886 . 10^6}{E} \tag{82}$$

Figure 28 shows that in cubic packagings of realistic dimensions the relative humidity is high, indicating that the risk of spoiling the product by growth of micro-organisms is great.

The properties used are $\pi = 180 \cdot 10^{-15}$ (for carton), $d = 0 \cdot 005$, and $E = 2 \cdot 88 \cdot 10^{-3}$. It is necessary to use a certain number of vent holes in the packaging of horticultural produce to prevent undesirable concentrations of oxygen, carbon dioxide or other volatiles. The vent holes serve as a ventilation system if CA-storage conditions are to be prevented. The mass transfer theory uses coefficient H which can be described with the situation of Fig. 4.4 and eqn (83).

$$H = \cfrac{1}{\cfrac{1}{H_{conv}} + \cfrac{1}{\xi \left(\dfrac{D}{d}\right)_{air} + (1 - \xi) \cdot \left(\dfrac{D}{d}\right)_{mat}}} \tag{83}$$

The overall transfer coefficient in packaging material is often given as a fraction of the diffusion coefficient in the system water vapour–air

$$D_{material} = \frac{D_{w-a}}{\mu} \tag{84}$$

The convective overall mass transfer coefficient is

$$H_{conv} = \frac{\alpha}{(\xi c)_{air}} \tag{85}$$

The flow of water through packaging materials can be presented with two equations

$$\dot{w} = \frac{\pi}{d} \cdot R_d \cdot T_K \cdot \Delta c \tag{86}$$

and

$$\dot{w} = \frac{D_{w-a}}{\mu d} \Delta c \tag{87}$$

The relationship between permeability π and diffusion resistance factor μ' is

$$\pi \cdot R_d \cdot T_K = \frac{D_{w-a}}{\mu'} \tag{88}$$

where $R_d = 462 \, J/(kg \cdot K)$ and $D_{w-a} = 2 \cdot 3 \cdot 10^{-5} \, m^2/s$.

Table 9 gives approximate permeabilities and diffusion resistance factors of some packaging materials. The mass transfer coefficient depends strongly on the percentage of vent holes in the packaging material. Table 10

TABLE 9
PERMEABILITY AND DIFFUSION RESISTANCE FACTOR OF PACKAGING MATERIALS

Material	π $(kg/(m.s.Pa).10^{-15})$	μ' $(\times 10^3)$
Cellulose acetate	63	2·9
Polypropene	0·34	536
Polyethene	0·17	1 070
Polyvinylchloride	1·7	107
Carton/paper	180	1·0

TABLE 10
MASS TRANSFER COEFFICIENT AS A FUNCTION OF VENT HOLE AREA AND PACKAGING MATERIAL THICKNESS

ξ (%)	$d = 1\,mm$	$d = 5\,mm$
30	$3·22.10^{-3}$	$1·12.10^{-3}$
10	$1·68.10^{-3}$	$4·31.10^{-4}$
5	$9·81.10^{-4}$	$2·26.10^{-4}$
1	$2·42.10^{-4}$	$5·01.10^{-5}$
0	$2·29.10^{-5}$	$4·60.10^{-6}$

shows calculated data. It is interesting that the difference between 0 and 1 % vent holes is a factor of 10.

5. COMPARISON BETWEEN THEORY AND EXPERIMENT

Three experiments are chosen for a comparison between theory and practice. Deviations will be found and an explanation for these deviations will be given leading to the applicability of analytical and numeric solution methods.

5.1. Box Filled with Cut Flowers
A carton box with the dimension $2X = 0·28$ m, $2Y = 0·48$ m and $2Z = 1·2$ m packed at $T_o = 19·6°C$, was exposed to cold air of $T_a = 5°C$ and $\varphi_a = 80\%$. The mean velocity of the air was 0·75 m/s. Figure 29 shows the measured course of temperature in the centre of the box as a function of time. The

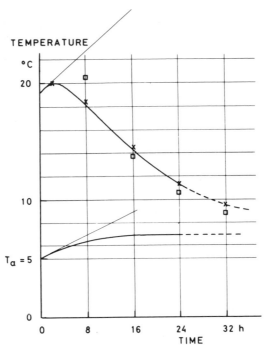

FIG. 29. Cooling and heating of a box of flowers: ——, measured centre temperature; x, six root solution; □, first root solution.

initial rise of the temperature can be explained by the heat generation of the cut flowers and the time lag for thermal diffusion. Initially, the temperature shows the adiabatic rise

$$\frac{\mathrm{d}T}{\mathrm{d}t} = \frac{Q}{\rho c_p} \tag{89}$$

The bulk density and specific heat of cut flowers (roses) are given in Table 5 and the heat production in Fig. 23. $Q = 74 \, \text{W/m}^3$ without transpiration, $\rho = 200 \, \text{kg/m}^3$ and $c_p = 4081 \, \text{J/(kg . K)}$, resulting in an adiabatic temperature rise of $0.32 \, \text{K/h}$. The calculated adiabatic temperature rise is in agreement with the measured initial course of temperature.

Within minutes after the start of the cooling process, the water vapour concentration is built up to pseudo-steady conditions, coming from almost saturated conditions in the box and forming a profile as drawn in Fig. 30. This profile is calculated by eqn (59), and the following physical properties: $E = 32.2 . 10^{-3} \, \text{l/s}$, $D = 11.3 . 10^{-6} \, \text{m}^2/\text{s}$, $H = 1.395 . 10^{-4} \, \text{m/s}$

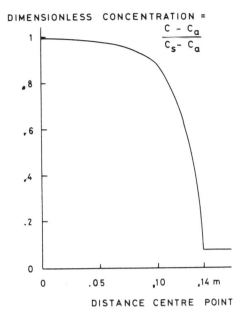

FIG. 30. Water vapour concentration in the X-direction of box with flowers.

and $X = 0.14$ m. The transpiration coefficient E and the bulk diffusion coefficient D are given in Table 8. The overall coefficient of mass transfer H is calculated for a packaging material thickness of $d = 5$ mm and diffusion resistance factor $\mu' = 1000$. The area of vent holes in the packaging is $\zeta = 3\%$.

The analytical solution demands a constant heat generation Q_{eff}, for the total temperature range. With eqn (77) and the calculated mean water vapour concentration, the effective heat generation is calculated. At steady state conditions the mean dimensionless water-vapour concentration deficit, using eqns (61) and (62), is

$$\frac{c_s - c}{c_s - c_a} = 0.045$$

Figure 31 shows the effective heat generation as a function of temperature for a set of mean dimensionless water vapour concentration deficits. The mean effective heat production in the temperature range of 20–5 °C is $Q_{eff} = 26$ W/m^3.

Introducing this calculated effective heat generation in the analytical solution (eqn (6)) gives an exact copy of the measurement. A heat

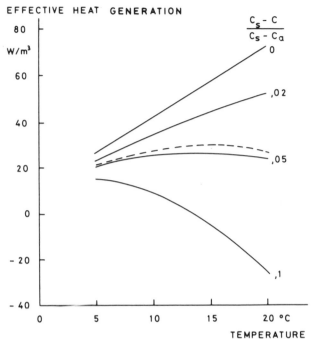

FIG. 31.   The effective heat generation of the flowers as a function of temperature
and concentration.

production of $Q = 74 \text{ W/m}^3$ gives no cooling effect, as Table 11 shows. This
example indicates the importance of the transpiration of the product for
predicting the cooling curve correctly.

The solution using only the first root instead of six roots is also close to
the measurement. Table 12 gives the data calculated with

$$\theta = j_c \cdot j_{ch} / 10^{t/f} \tag{90}$$

As can be expected, the first root solution is not correct in the beginning
of the cooling process (at low Fo numbers). Furthermore, eqn (39) shows
the importance of using the correct definition of the dimensionless
temperature. For the first root solution, the dimensionless temperature is
based on the difference between initial product temperature and the final
product temperature. The final temperature is calculated with eqn (49)

$$\frac{T_e - T_a}{T_o - T_a} = 0 \cdot 144 \tag{91}$$

TABLE 11

CALCULATED CENTRE TEMPERATURE IN A BOX
FILLED WITH CUT FLOWERS

Time (h)	Centre temperature $Q_{eff} = 26$ (°C)	$Q_{eff} = 74$ (°C)
0	19·6	19·6
8	18·4	23·8
16	14·3	23·0
24	11·5	22·1
32	9·7	21·8
64	7·3	23·7
100	6·9	27·7

TABLE 12

CALCULATION OF THE CENTRE TEMPERATURE
OF A BOX WITH CUT FLOWERS USING THE FIRST
ROOT SOLUTION

Time (h)	Temperature (°C)
8	20·5
16	13·9
24	10·7
32	9·0

$f = 26·3$ (eqn (21))
$j_c = 2·032$ (eqn (22))
$j_{ch} = 1·064\,5$ (eqn (42))

where

$$\mathrm{Po}_x = \frac{QX^2}{\lambda(T_o - T_a)} = \frac{26 \times 0·14^2}{0·136 \times (19·6 - 5)} = 0·257$$

$$\mathrm{Bi}_x = \frac{\alpha X}{\lambda} = \frac{8·7 \times 0·14}{0·136} = 8·96$$

$$N = 2·17 \qquad \text{(see Fig. 10)}$$

The final product temperature is

$$T_e = T_a + \theta_{ce}(T_o - T_a) = 5 + 0·144(19·6 - 5) = 7·1\,°C \qquad (92)$$

The calculated product end temperature is equal to the measured end temperature of another experiment. In this experiment cut flowers at 5 °C were put in the box and the box was stored in the chill room at 5 °C and 80 % relative humidity. After 16 h the product end temperature was reached. The initial temperature rise during this experiment (see Fig. 29) was 1·5 K/(8 h). The heat generation at 5 °C of the cut flowers was

$$Q = \rho c_p \frac{dT}{dt} = 200 \times 4081 \times \frac{1\cdot5}{8 \times 3600} = 43 \text{ W/m}^3$$

This is too high, for the heat production at 5 °C is 26 W/m³ as a maximum. According to eqn (35) there is a relation between residual centre temperature and cooling time

$$(T_e - T_a)_c = \frac{Q}{b\lambda} f_x a \tag{93}$$

The constant b is given in Table 2 for the basic geometries, and can be chosen between the value of a cube and a square rod, so b = 1·6. The real value of b is 1·63. This follows from $b = f_x/(f_{xyz} . N)$. The correlation between residual temperature difference and cooling time f_x is poor.

$$2\cdot1 \approx \frac{20}{1\cdot63 \times 0\cdot136} \times 1\cdot66 \times 10^{-7} \times 26\cdot3 \times 3600 = 1\cdot42$$

5.2. Pallet of Carton Boxes Filled with Pears

The dimension of the pallet stack was $2X = 1\cdot0$ m, $2Y = 1\cdot2$ m and $2Z = 1\cdot5$ m. The ambient air temperature during cooling was 4·5 °C and the

TABLE 13

CALCULATED TEMPERATURE OF THE CENTRE
TEMPERATURE OF A PALLET STACK OF PEARS
WITH THE FIRST ROOT SOLUTION

Time (h)	Dimensionless temperature	Temperature (°C)
0	2·03	25·0
50	1·256	17·2
100	0·777	12·3
150	0·480 8	9·3
200	0·297 4	7·5

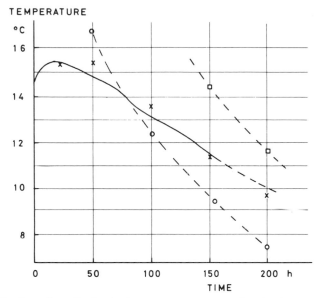

FIG. 32. Cooling of a pallet load of pears: ——, measured centre temperature: x, six root solution ($Q_{eff} = 5$ W/m^3); $\bigcirc$, first root solution ($Q_{eff} = 0$ W/m^3); $\square$, first root solution ($Q_{eff} = 25$ W/m^3).

product initial temperature $T_o = 14.6\,^\circ$C. The side walls of the pallet stack were covered with plastic shrink film.

The first root solution without heat production leads to $f = 240$ h (eqn (21)) and $j_c = 2.03$ (eqn (22)) when the heat transfer coefficient is $\alpha = 7.36$ W/(m^2.K) and $v = 0.4$ m/s. The calculated points are given in Fig. 32, which are also given in Table 13 using 4.5 °C as the final product temperature. This calculation indicates that the effective heat generation is much higher than zero.

The effective heat production as a function of temperature for the pears in the pallet stack is given in Fig. 33. The dimensionless water vapour concentration deficit is

$$\frac{c_s - c}{c_s - c_a} = 0.179$$

using the following properties listed in Table 14. The effective heat generation under these circumstances is 40 W/m^3 at 14.6 °C, and 10 W/m^3 at 5 °C, resulting in a mean effective heat generation of 25 W/m^3. In Fig. 32 the first root solution is also given; $Q_{eff} = 25$ W/m^3. It is clear that this

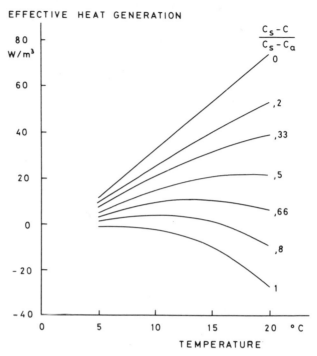

FIG. 33.   Effective heat generation of packed pears as a function of temperature
and dimensionless concentration (0·179 when calculated).

analytical solution can never give correct results because of the fact that the
first root solution gives correct results at high Fo numbers or after a long
time or if the dimensionless temperature $\theta < 0·6$. The calculated product
end temperature, if $Q_{eff} = 10 \, W/m^3$, is $T_e = 7·5 \, °C$.

The complete solution of Luikov, with six roots, gives no cooling effect if
$Q_{eff} = 25 \, W/m^3$, while the best fit gives $Q_{eff} = 5 \, W/m^3$, which indicates that

TABLE 14

In the direction	X	Y	Z
ξ = vent hole area (%)	1	1	25
D = diffusion coeff.	$5·04.10^{-6}$	$5·04.10^{-6}$	$5·04.10^{-6}$
H = overall mass transfer coeff.	$2·55.10^{-3}$	$2·55.10^{-3}$	$5·45.10^{-3}$
d = thickness (mm)	0·05	0·05	0·05
μ = permeability	$1·070.10^6$	$1·070.10^6$	$1·070.10^6$

EFFECTIVE   HEAT   GENERATION

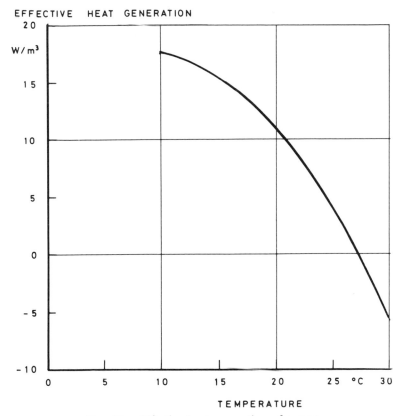

FIG. 34.   Effective heat generation of potato.

the water vapour concentration calculation must be wrong. Only the last point is in agreement with the experiment: the initial temperature rise is $\Delta T/\Delta t = 4 \cdot 5$ K (50 h) resulting in a heat generation of $Q = 56$ W/m^3, which is in accordance with the heat generation at 15 °C and very high relative humidity.

5.3. Potatoes in a Metal Box Without Vent Holes

The dimensions of the metal box[12] are $2X = 0 \cdot 455$ m, $2Y = 0 \cdot 715$ m and $2Z = 0 \cdot 715$ m. The walls of the box were kept at a temperature of $T_a = 19 \cdot 9$ °C and the product start temperature was 29·5 °C. The walls of the box were below the dew point of the air in the box, resulting in a wet-wall surface. The water vapour concentration c_a is calculated at 19·9 °C and $\zeta = 100$ %, which allows us to calculate the effective heat generation of the

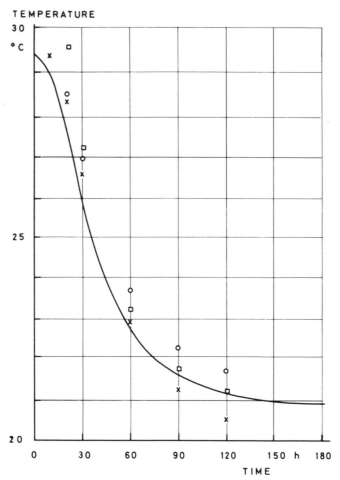

FIG. 35. Cooling of potatoes: ——, measured centre temperature; x, six root solution; ○, six root solution using the residual temperature of the product centre as the cooling temperature; □ first root solution.

potatoes in this box for the temperature range during cooling (see Fig. 34). The mean heat generation is $Q_{eff} = 2.5$ W/m^3 during the cooling process. The solution of Luikov with six roots leads to reasonable results using $\alpha = 5.43$ W/(m^2.K) (Fig. 35). The tail of the calculated cooling curve is lower than the measured curve, because six roots are not enough at high Fo numbers or at low dimensionless temperatures. If the residual temperature of the product centre is used as the cooling temperature, the six root Luikov

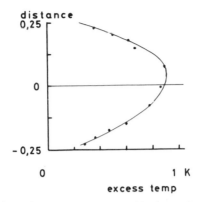

FIG. 36. Shift of maximum temperature. Horizontal scale division 0·2 K.

solution then gives cooling times which are too long. The residual temperature is equal to the measured product end temperature

$$Po_x = \frac{QX^2}{\lambda(T_o - T_a)} = \frac{12 \times 0·227^2}{0·311 \times 9·6} = 0·207$$

$$Bi_x = \frac{\alpha x}{\lambda} = \frac{5·43 \times 0·227}{0·311} = 3·963$$

$$N = 3·70$$

$$\frac{T_e - T_a}{T_o - T_a} = 0·1235 \qquad \text{giving} \qquad T_e = 20·7\,°C$$

The first root solution with $T_e = 20·7\,°C$ gives very good results compared to the six root Luikov solution. The next variables are calculated for the cooling of the potatoes: $f = 72·7$ h, $j_c = 1·9168$, $j_{ch} = 1·00395$.

The natural convection has an important effect on the place of the maximum temperature. It is shifted from the centre to 5 cm higher (Fig. 36). However, the maximum temperature is correctly predicted by eqn (55). The accuracy of the predicting methods can be evaluated against the consequences of a deviation and the degree of complication necessary for the calculation. The aspects of quality maintenance and refrigeration power have to be treated separately. Table 15 gives a comparison for increasing degrees of complication.

As the real cooling time for the thermal centre is longer than the exponential cooling time, quality deterioration during cooling is underestimated by only looking for the f-factor only. Consideration of the

TABLE 15

COMPARISON OF PREDICTING METHODS FOR COOLING TIMES AND KEEPING CONDITIONS IN ORDER OF INCREASING COMPLICATIONS AND TIME CONSUMPTION

	Factor	Formula	Consequences for quality	refrigeration power
Cooling time	f	$\dfrac{X}{a}\left(1+\dfrac{2}{Bi}\right)$	Optimistic	Too small
		f_{xyz}		
	$f.j$	Graph	Slightly optimistic	Close to reality, difficult to determine
	j_H	$\dfrac{1+\dfrac{Po}{2\cdot303}\dfrac{a}{X^2}f_{xyz}}{1-\dfrac{Po}{N}\left(1+\dfrac{2}{Bi}\right)}$	Pessimistic	Close to reality
Keeping conditions	θ_e	$\dfrac{Q_{T+ve}X^2}{N\lambda}\left(1+\dfrac{2}{Bi}\right)$	Pessimistic if reasonable evaporation	Close to reality

lag factors gives a better estimate; however, the heating factor is less yielding because of the complication by incorporating the shape factor N, which must be read from a graph. The same holds for the estimation of the residual temperature.

Computerised calculations are, because of their singularity, not really valuable for practical use if they are not transformed to easily readable nomograms.

In many cases the available refrigeration power is not adjusted to the maximum which could be absorbed in a short initial period. This means that even the exponential cooling process cannot be realised. If the refrigerating power is nearly constant the cooling time can be estimated simply by

$$\Delta t = \frac{m \cdot c_p}{Q_{refr.} \cdot \Delta T} \tag{94}$$

For quality conservation this approach yields an over estimate.

The six root (Luikov) and the first root (Pflug) analytical solutions are good methods to estimate the cooling curve of the product centre temperature of unit loads cooled with air flowing around the load. Under the following conditions the calculation will give a fairly good approximation of the cooling process:

1. *Low air velocity* if vent holes are present in the packaging. High air velocities around the unit load introduce fluctuating pressure differences around the load. This effect results in small air velocities in the unit load. Convection now supports the diffusion of mass and heat in such a manner that the analytical solutions behave badly without correction of the diffusion properties.

2. *Constant heat generation* for the cooling period. Analytical solutions use, in most cases, constant heat generation. Without transpiration of the products this condition never holds, resulting in very poor results, as demonstrated with the cooling of pears. It indicates that the calculation of the water vapour concentration deficit is the determining factor in the calculation of the effective heat generation. The six root (Luikov) solution for the potato shows that, with constant mean heat generation, the result is not good enough. In the beginning the real effective heat generation is negative which produces very fast cooling. At the end of the cooling process the real effective heat generation is high ($15 \, W/m^3$) compared to the mean effective heat generation.

3. *More than one packaging in the stack.* Calculation of the water vapour concentration deficit under these practical conditions is nearly impossible. This yields, as in the example of the pears, poor estimates of the effective heat generation.

Every calculation method, analytical or numerical, will give poor results if the evaporation of the product is not incorporated in the solution.

REFERENCES

1. LUIKOV, A. V., *Analytical Heat Diffusion Theory*, 2nd edn, Academic Press, New York, 1965.
2. DALGLEISH, N. and EDE, A. J., Charts of determining centre, surface and mean temperature in regular geometric solids during heating or cooling, Report No. 192, National Engineering Lab., Glasgow, 1965.
3. PFLUG, I. J. and KOPELMAN, I. J., Correlating and predicting transient heat transfer rates in food products, Annexe 1966-2, Trondheim, IIF, Paris, 1966.
4. MEFFERT, H. F. TH., A new chart for the solution of transient heat transfer problems, Annexe 1970-1, London, IIF, Paris, 1970.
5. MEFFERT, H. F. TH., The relation between cooling rate and equilibrium excess temperature of unit loads of agricultural produce, Annexe 1972-1, Freudenstadt, IIF, Paris, 1972.

6. SAINSBURY, G. F., Cooling apples and pears in storage rooms, Marketing Research Report No. 474, USDA, Washington DC, 1961.
7. LINDAU, E. H. and MATTHISEN, I. V., Transportation and ripening of palletized bananas, D3-14, XVth Intern. Congress of Refrigeration, Venice, 1979.
8. TCHUMAK, I. G., MURASKOV, V. S. and PETROVSKY, V. P., Heat and moisture exchange in fruit storage rooms, Annexe 1970–3, IIF, Paris, 1970.
9. FOCKENS, F. H. and MEFFERT, H. F. TH., Biophysical properties of horticultural products as related to loss of moisture during cooling down, *J. Sci. Fd Agric.*, 1972, **23**, 285–98.
10. COWELL, N. D., Some aspects of the loss of weight from fruit in cool storage, Thesis, University of New South Wales, 1960.
11. RUDOLPHIJ, J. W., VERBEEK, W. and FOCKENS, F. H., *Lebensm. Wiss. u. Technol.*, 1977, **10**, 153–8.
12. BEUKEMA, H. J., Heat and mass transfer during cooling and storage of agricultural products as influenced by natural convection, Thesis, University of Wageningen, Pudoc Wageningen, 1980.
13. GRÖBER, H., ERK, S. and GRIGULL, U., *Fundamentals of Heat Transfer*, McGraw Hill, New York, 1961.
14. GRIGULL, U., Näherungslösungen der nichtstatiönaren Wärmeleitung, *Forschungs–Ingenieurswesen*, 1966, **32**, 11–18.

Chapter 3

THE PREPARATION OF FRUIT JUICE SEMI-CONCENTRATES BY REVERSE OSMOSIS

M. Demeczky, M. Khell-Wicklein and E. Godek-Kerék

Central Food Research Institute,
Budapest, Hungary

SUMMARY

On the basis of data quoted in the literature, investigations and results obtained by the use of the reverse osmosis technique in fruit juice concentration are summarised.

Experiments carried out by the authors using pilot equipment with cellulose acetate membranes verified that only semi-concentrates of 30–35% solids content can be produced using the reverse osmosis method. The high osmotic pressure of the fruit and vegetable juices, the present technical conditions, compaction of the membranes, decrease in flux, losses of the main chemical compounds and particularly of aroma components limit the use of higher concentration rates. However, semi-concentrates produced by reverse osmosis are of high quality and are better than those prepared by vacuum evaporation.

Experimental data showed that most of the semi-concentrates can be stored at room temperature for only a few weeks, but products of high acid-content at $+3°C$ can be stored for 12 months.

A preliminary analysis of the economics and energy requirements to be expected is presented. Reliable results for these important factors can only be based on long-term experiments on a large scale.

1. INTRODUCTION

The recognition of the disproportionality between the growth of world population and food production urges the food experts to find new ways to utilise food raw materials and to preserve their nutrient contents to the fullest possible extent. The energy crisis of the past couple of years resulted in higher food prices in the developed countries and hindered the development of the food industry in the developing world (where population growth is the fastest), because the energy consumption of modern food industry is high. These factors led to the revision of existing technologies and the development of new, low-energy, no-waste, processing technologies.

Based on their success in water purification, the advantage of membrane separation processes was recognised early by food experts. An additional advantage of membrane separations in food industry applications is that the concentration of heat-sensitive biological materials, recovery of valuable nutrients from by-products and separation of liquid mixtures into their constituents became possible without heat input.

Research activity, directed at the possible applications of membrane separations in the food industry, has been very intense since the late 1960s. This research was focussed mainly on ultrafiltration and reverse osmosis, the replacement of certain elements of existing technologies and the development of new technologies based on the membrane separation processes. Soon after the first results of this research became known, the examination of the adaptability of existing water-treatment membranes and apparatus also began. This was closely followed by the development of specific apparatus designed for the purposes of the food industry, e.g. DDS, PCI, etc.

Though membrane separations belong to up-to-date food processing techniques and will remain one of them, their general use is severely hindered by a number of problems yet to be solved, e.g. the price of the apparatus, the permeability, selectivity, price and cleaning, sterilization of the membranes, etc. The most advanced member of the family is ultrafiltration, a routine process of the modern dairy industry.

On the other hand reverse osmosis is only put into practice in water treatment[1,2] and in the dairy industry.[3-6] Elsewhere this technique has not exceeded the laboratory or the pilot-plant level in the food industry.

There is an ever growing demand for soft drinks, and especially fruit and vegetable drinks[7] which is a very healthy trend. Therefore, it is essential that the valuable nutrients of the raw juices be preserved during concentration.

Also, the quality of the concentrates produced significantly influences their market value.

Concentration of fruit and vegetable juices by reverse osmosis has been studied since the early 1960s. Morgan *et al.*[8] reported their work first. They produced orange and apple concentrates with 40% solids content using cellulose acetate membranes. At apple juice concentrated to 30° Brix the permeability of the membrane was $164 \, dm^3 \, m^{-2} \, d^{-1}$ and the solids content of the permeate was less than 1%. They examined the changes of the aroma values of the orange and apple concentrates as well, because these changes profoundly influence the value of the concentrate. It is well known that the number of aroma components and their ratio changes with fruit variety and ripeness. They concluded that the higher the number and concentration of low molecular weight, water-soluble components in the raw juice, the higher the processing loss in reverse osmosis. The concentrate was judged excellent by tasting panels. In the chromatogram of the permeate all the components present in the original raw juice of Delicious apple could be identified. The cellulose acetate membranes retained the less water-soluble components of orange aroma to a larger extent, since these compounds occurred mostly in the oil emulsion.[9]

Gherardi *et al.*[10] obtained similar results with model solutions of saccharose and citric acid and with apple, grapefruit and orange juices. They used cellulose acetate membranes and an apparatus produced by Abcor Inc. Changes of the permeability of the membrane, and its selectivity towards the major inorganic and organic constituents of the juices, were examined by comparing the composition of the raw juice, permeate and concentrate. The solids content of the apple concentrate was 25·2% while those of the orange and grapefruit concentrates were 24% and 27·1%, respectively. The total losses of the concentration process were as follows: 16·7% acids and 3·2% directly reducing carbohydrates for apple juice; 1·05% acids and 2·25% carbohydrates for grapefruit juice and 2·8% acids and 7·6% carbohydrates for orange juice. According to their model experiments with saccharose and citric acid there was almost no loss in these component up to a solids level of 23–25%, while above this level the permeate contained increasing amounts of these components. This was attributed to the saturation of the membrane. The loss of the essential amino acids and mineral salts was considered acceptable. The selectivity of the membrane for ascorbic acid was emphasised. Their examinations indicated that the transfer of the low molecular weight aroma components through the cellulose acetate membrane was higher at the beginning of the concentration process. This observation is of high technological importance because this fraction of the permeate can be either recycled into the raw juice in the final stage of the concentration operation, or it can be concentrated separately by a more suitable membrane and then mixed with the final concentrate. The appearance of the low molecular weight alcohols, esters, etc. in the permeate had no detrimental effect on the quality of the concentrates which, after dilution to the solids content of the original juices, retained the characteristic taste of fresh juices.

During the comparison of type DDS M-990, M-890 and Abcor AS-197 and AS-187 membranes,[11] the productivity of the DDS membranes was found higher for orange juice. With a raw orange juice of 11° Brix the DDS membranes displayed $400\,dm^3\,m^{-2}\,d^{-1}$ permeability. With raw juices of 20° Brix this value was $150\,dm^3\,m^{-2}\,d^{-1}$. Under identical conditions the values obtained with the Abcor membranes decreased from $200\,dm^3\,m^{-2}\,d^{-1}$ to less than $100\,dm^3\,m^{-2}\,d^{-1}$. Higher losses were obtained over 20° Brix concentration levels. The decrease of the acid content of the semi-concentrate with respect to the original value was advantageous. The loss of the productivity of the membranes could be explained by pectin deposition preventing the transfer of water molecules through the membrane.[12]

Bolin and Salunkhe[13] reported the results of their comparative investigation dealing with the quality of apple, peach and sour cherry concentrates produced by freeze drying, evaporation, osmosis and reverse osmosis. Using 6·8–8·2 MPa pressure for reverse osmosis and cellulose acetate membranes, the solids-content levels changed from 12 to 32 % for apple juice, from 14 to 29 % for peach juice and from 16 % to 32 % for sour cherry juice. The permeates had a solids content of 0·1–0·5 %, attributed to the low selectivity of the membrane towards the organic acids and anthocyanins. They compared the aromagrams of the raw juices and the concentrates diluted to the original solids content level and observed that the amount of almost all the aroma component decreased, and the extent of the decrease was significantly higher for the apple juice than for the other two juices. Benzaldehyde could not be detected at all in the sour cherry concentrate. The tasting panels could not detect that change. Gas chromatographic separations proved that freeze drying preserved the total aroma content of fruit juices best, followed by reverse osmosis, pervaporation and osmosis. Tasting could detect differences between the original and concentrated–rediluted juices only in the case of osmotic concentration of apple and peach juices.

In another comparative investigation[14] semi-concentrates of 30 % solids content by refractive index were produced by thermal evaporation (the apparatus was also equipped with an aroma recovery unit), freeze drying and reverse osmosis. Again, freeze drying was rated best and thermal evaporation the worst. The authors concluded that reverse osmosis was best restricted for the concentration of less aromatic fruit juices. Only those membranes known at the time of the investigation could be used for the concentration of more aromatic juices which had permeabilities in the $25–160\,dm^3\,m^{-2}\,d^{-1}$ range.

Pompei and Rho[15] investigated the use of DDS M-990, M-950 and M-900 membranes for the concentration of the juice of *Passiflora edulis*. The permeability of the membranes was in the $150–230\,dm^3\,m^{-2}\,d^{-1}$ range for juices of 16 % solids content. This value decreased to $45–50\,dm^3\,m^{-2}\,d^{-1}$ for juices of 26 % refractory content. However, due to their selectivity these membranes proved suitable for the concentration process. Therefore, the authors suggested the use of the process at pressures above $7\,MPa$ and optimised flow conditions for the large-scale concentration of the juice.

Grape concentrates are used not only for soft drinks but also for wine making (no wonder that most publications come from this field). Apart from reporting the production and testing of grape concentrates several authors[16–20] have reported on the use of reverse osmosis, ultrafiltration and electrodialysis in wine making.[19–24] This indicates the versatility of membrane separations even within the framework of a single branch of the industry.

The primary goal of the experiments was not so much the production of grape soft drinks, rather the increase of the sugar content and the decrease of the acid content of low-grade musts to be used in wine making. The Hungarian Wine Act of 1977 prohibits the use of sugar and the depression of the acid content in wine by $CaCO_3$, so only physical methods can be used for the regulation of the carbohydrate–acid ratio. Evaporation cannot be used for this purpose, but reverse osmosis seems very promising. Both requirements can be fulfilled in a single step by reverse osmosis. The carbohydrate concentration can be increased by 1–2 % by preconcentrating the total volume of the must or by the introduction of grape juice concentrate. The first method was used in a pilot-plant scale experiment and the carbohydrate content could be increased by a degree equivalent to 2 % alcohol, without the occurrence of the 'cooked' taste and formation of hydroxy methyl furfurol.[15]

Wucherfennig[19] pointed out in the discussion section of his paper that though grape juices could be concentrated by reverse osmosis the productivity of the membranes decreased significantly during the concentration process.

2. PREPARATION OF FRUIT AND VEGETABLE SEMI-CONCENTRATES

The development of reverse osmosis allows for the extension of fruit and vegetable juice concentration experiments, so that processing can be carried out at the place of fruit production and that the quality of the products can be improved. The technological experiences gained in the course of these experiments, along with the literature data, prepare the foundation for future full-scale processing technologies. In the future further energy consumption measurements and economical calculations have to be carried out as well.

2.1. Raw Materials Used for the Experiments
Sour cherry, red currant, blackcurrant, strawberry, peach and apple juices, the major domestic sources of soft drinks produced in Hungary, were included in the experiments. Attempts were made to concentrate carrot juice, red beet juice and tomato serum as well.

2.2. Preparation and Pretreatment of the Juices
After the usual washing, pitting and crushing operations of the canning industry the experimental materials were treated with a pectin-decomposing enzyme, 'Phylazine-10'. This treatment was followed by sedimentation and filtration. Decomposition of the pectin content increased the juice yields, but its major role was to decrease the amount of gel-like materials and moderate their precipitation on the membranes. Prior to concentration the juices were filtered to a crystal-clear state.

2.3. Apparatus and Membranes Used in the Concentration Experiments
A laboratory-scale, flat-membrane apparatus with a $0 \cdot 1 \, m^2$ useful filtration cross section was used for the concentration of $10-15 \, dm^3$ juices. The upper pressure limit of the system was $6 \cdot 0$ MPa, its membrane pump could deliver $20 \, dm^3 \, h^{-1}$ liquid.

A similarly designed, but larger apparatus with a membrane surface of $0 \cdot 9 \, m^2$ was used for the concentration of larger juice volumes ($15-50 \, dm^3$). A $300 \, dm^3 \, h^{-1}$ capacity piston pump was used to feed the juices

A back-pressure regulator was used in both systems to control the liquid pressure. The concentrate flow was cooled to prevent any increase of the temperature of the juices during concentration.

The experimental cellulose acetate membranes marked 73/14, 73/28, 73/29, 77/9, 78/19, 80/5 and 80/6 were prepared in our laboratory from a

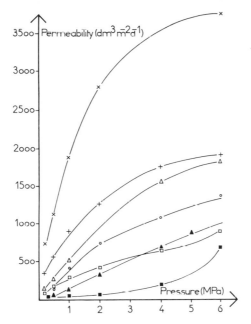

FIG. 1. Water permeability of the membranes as a function of pressure.
Membranes: $\times = 73/14$; $\triangle = 73/28$; $\square = 73/29$; $+ = 77/9$; $\blacktriangle = 73/19$; $\bigcirc = 80/5$;
$\blacksquare = 80/6$.

three-component casting mixture (cellulose acetate, acetone and for-
mamide). The water permeability of the membranes as a function of
pressure is shown in Fig. 1. Their selectivity, measured by a 5% glucose
solution were as follows: 33, 100, 100, 80, 92, 98 and 87%.

2.4. Vacuum Concentrator
For comparison purposes concentrates were also produced from certain
juices by a vacuum concentrator. Juice (40 dm^3) was fed into the
concentrator operated at 50 °C, 92 kPa for 40–60 min.

2.5. Fruit Juice Semi-Concentrates
The solids content of fruit juice concentrates is generally 65–68%.
Superconcentrates can be produced from these concentrates by further
water removal, their solids content is higher than 70%. Concentrates of
25–45% solids content are to be considered semi-concentrates. Their
osmotic pressure is lower than 9·5 MPa.

TABLE 1

CONCENTRATION DATA OF SOUR CHERRY SEMI-CONCENTRATES PRODUCED BY REVERSE OSMOSIS AND BY VACUUM EVAPORATION

Constituents and properties examined	Juice	Reverse osmosis		Evaporation	Concentration (concentrates diluted to the refractory % of juice)			
		Concentrate	Losses %	Concentrate	Reverse osmosis		Evaporation	
					Absolute values	Differences	Absolute values	Differences
Solids content (refractory %)	16·0	30·0	5·3	38·0	16·0	0	16·0	0
pH value	3·1	3·1	—	3·4	No data		No data	
Density (kg.dm^{-3})	1·069	1·135	—	1·172	No data		No data	
Acid content (calculated as citric acid g.dm^{-3})	14·6	24·5	14·7	36·3	12·3	−2·3	13·9	−0·7
Reducing sugars (calculated as glucose g.dm^{-3})	100·1	211·5	0	268·7	106·2	+6·1	103·2	+3·1

Pressure applied 6·0 MPa; pressure difference $\Delta P = 3·6 \to 0·4$ MPa (ΔP = pressure applied – approximative osmotic pressure); permeability average 87 dm^3 m^{-2} d^{-1} (145 → 60); rate of concentration 2·11; membrane 73/29; losses (%) calculated on the basis of the concentrate.

2.5.1. Sour Cherry

Sour cherries of two successive years were used for the experiments. The raw material of the second experiment series was a less aromatic, 'empty tasting' fruit.

Compared with the other fruit juices sour cherry juices contained more carbohydrates. Therefore, owing to the initially high osmotic pressure of the sour cherry juices, a 2·6-fold concentration could be achieved at 6·0 MPa operating pressure (Tables 1 and 2). The permeates produced were practically colourless and contained the characteristic aroma components of sour cherry. Depending on the characteristics of the membranes used the losses were varied; the acid losses were significant (14·7–23·6 %) while the losses of carbohydrates and other components were much lower.

There was no significant difference between the colour, taste and smell of the first fresh juice and the 30 % refractory level semi-concentrate rediluted to the original solids content. When the rediluted samples prepared by reverse osmosis and vacuum evaporation from an identical raw juice were compared, the taste and smell of the sample obtained by vacuum evaporation proved considerably inferior, but there was no difference between the colour of the samples.

In the case of the second raw juice the rediluted semi-concentrate was judged 'characterless', therefore significantly inferior to the original juice. There was no difference between the semi-concentrates produced by the two different methods. The explanation of the unfavourable result lies probably in the fact that lower grade raw fruit was used and both the acid and carbohydrate losses were larger. The losses indicated that the low molecular weight aroma components were also lost in part.

The permeability of the membranes decreased significantly as the concentration increased (Fig. 2). This was due to the gradual decrease of the pressure difference (ΔP) between the concentration pressure and the osmotic pressure of the concentrate. The permeabilities relating to the overall process were practically identical in the two experiments (87 and 83 $dm^3 m^{-2} d^{-1}$).

2.5.2. Peach Juice

During concentration the solids content of peach juice could be increased from 11 % to 30·7 % (Table 3). The permeate was colourless and contained fragrant components. 13·8 % of the acid and 4·1 % of the carbohydrate content of the juice permeated the membrane. The average permeability was 145 $dm^3 m^{-2} d^{-1}$ at a concentration ratio of 3·1. Double-blind taste determinations proved that the fresh juice was significantly better than the

TABLE 2
CONCENTRATION DATA OF SOUR CHERRY SEMI-CONCENTRATES PRODUCED BY REVERSE OSMOSIS AND BY VACUUM EVAPORATION

Constituents and properties examined	Juice	Reverse osmosis			Evaporation	Concentration (concentrates diluted to the refractory % of juice)			
						Reverse osmosis		Evaporation	
		Concentrate	Permeate	Losses %	Concentrate	Absolute values	Differences	Absolute values	Differences
Solids content (refractory %)	16·9	31·0	2·0	5·8	31·0	16·9	0	16·9	0
pH value	3·2	3·3	3·1	—	3·2	No data		No data	
Density (kg.dm⁻³)	1·064	1·127	1·007	—	1·127	No data		No data	
Acid content (calculated as citric acid g.dm⁻³)	13·7	20·5	6·3	23·6	27·1	10·5	−3·2	13·9	+0·2
Reducing sugars (calculated as glucose g.dm⁻³)	113·6	231·8	9·6	4·3	226·6	119·3	+5·7	116·6	+3·0
Total polyphenol content (mg.dm⁻³)	2 147	3 576	39	0·9	3 414	1 840	−307	1 757	−390
Amino nitrogen content (gm.dm⁻³)	461	964	75	8·4	964	496	+35	496	+35
AA + DAA content (mg.kg⁻¹)	63·5	104·4	0	0	91·7	56·9	−6·6	50·0	−13·5

Pressure applied 6·0 MPa; pressure difference $\Delta P = 3·4 \rightarrow 0·2$ MPa (ΔP = pressure applied − approximative osmotic pressure); permeability average 83 dm³ m⁻² d⁻¹ (446 → 35); rate of concentration 2·6; membrane 80/5; losses (%) calculated on the basis of the permeate.

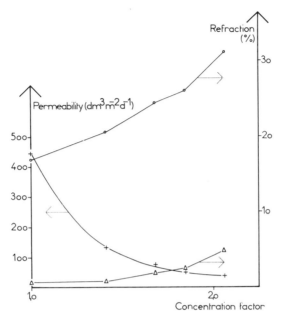

FIG. 2.   Water permeability of the membrane and the solids content of concentrate and permeate as a function of the concentration factor of sour cherry juice. Membrane = 80/5; pressure = 6·0–6·4 MPa; + = water permeability; ○ = solids content of the concentrate; △ = solids content of the permeate.

rediluted semi-concentrate. The colour of the rediluted preconcentrate produced by vacuum evaporation was judged significantly worse by the panelists than that of the reverse-osmosis treated sample. The two products did not differ in smell or taste.

The 78 dm^3 m^{-2} d^{-1} value obtained at a refractory level of 30·7 % was considerably higher than the values obtained by the 73/29 membrane for other fruits.

2.5.3. Red Currant Juice

Concentration was carried out in two stages. The permeate of 1·6 % refractory content corresponding to the 29·9 % refractory content semi-concentrate was recirculated at the end of the concentration step. The final semi-concentrate of 27 % solids content shown in Table 4 was obtained by mixing the first and second concentrates, so it contained the components reclaimed from the first permeate as well. The slightly pink permeate contained 7·6 g dm^{-3} carbohydrates and 4·8 g dm^{-3} acids. The solids

TABLE 3

CONCENTRATION DATA OF PEACH JUICE SEMI-CONCENTRATES PRODUCED BY REVERSE OSMOSIS AND BY VACUUM EVAPORATION

Constituents and properties examined	Juice	Reverse osmosis			Evaporation	Concentration (concentrates diluted to the refractory % of juice)			
						Reverse osmosis		Evaporation	
		Concentrate	Permeate	Losses %	Concentrate	Absolute values	Differences	Absolute values	Differences
Solids content (refractory %)	11·0	30·7	0·5	3·0	27·5	11·0	0	11·0	0
pH value	4·3	4·1	4·1	—	4·1	No data		No data	
Density at 17°C (kg.dm⁻³)	1·046	1·122	1·009	—	1·110	No data		No data	
Acid content (calculated as citric acid g.dm⁻³)	2·8	7·7	0·6	13·8	7·2	2·6	−0·2	2·7	−0·1
Reducing sugars (calculated as glucose g.dm⁻³)	38·7	94·2	2·3	4·1	107·2	35·5	−3·2	40·4	+1·7

Pressure applied 6·0 MPa; pressure difference $\Delta P = 5·4 \to 0·4$ ($\Delta P =$ pressure applied − approximative osmotic pressure); permeability average 145 dm³ m⁻² d⁻¹ (208 → 78); rate of concentration 3·1; membrane 73/29; losses (%) calculated on the basis of the permeate.

TABLE 4
CONCENTRATION DATA OF RED CURRANT SEMI-CONCENTRATES PRODUCED BY REVERSE OSMOSIS AND BY VACUUM EVAPORATION

Constituents and properties examined	Juice	Reverse osmosis			Evaporation	Concentration (concentrates diluted to the refractory % of juice)			
						Reverse osmosis		Evaporation	
		Concentrate	Permeate	Losses %	Concentrate	Absolute values	Differences	Absolute values	Differences
Solids content (refractory %)	8·5	27·0	0·4	3·3	28·2	8·5	0	8·5	0
pH value	3·0	2·8	3·2	—	2·9	No data	No data	No data	No data
Density (kg. dm⁻³)	1·035	1·122	0·988	—	1·126				
Acid content (calculated as citric acid g. dm⁻³)	19·2	63·4	1·4	5·3	69·6	18·4	−0·7	19·3	+0·1
Reducing sugars (calculated as glucose g. dm⁻³)	50·2	182·9	1·6	2·3	184·9	53·1	+2·9	51·2	+1·0
Total polyphenol content (mg. dm⁻³)	581	2075	19	2·4	2179	603	+22	604	+23
Amino nitrogen content (mg. dm⁻³)	192	521	69	25·8	548	151	−41	152	−40
DAA content (mg. kg⁻¹)	139·9	225·9	24·3	12·1	458·9	71·2	−68·7	138·1	−1·8
AA content (mg. kg⁻¹)	120·3	425·4	16·7	9·6	176·3	134·0	+13·7	53·1	−57·2
AA + DAA content (mg. kg⁻¹)	260·2	651·3	41·0	11·0	635·2	205·2	−55	191·2	−69

Pressure applied 6·1 MPa; pressure difference $\Delta P = 4·9 \rightarrow 1·3$ (ΔP = pressure applied − approximative osmotic pressure); permeability average 112 dm³ m⁻² d⁻¹ (325→41); rate of concentration 3·56; membrane 80/6; losses (%) calculated on the basis of the permeate.

content of the so-called second permeate was considerably lower; its refractory value was 0·4 %, carbohydrate content was 1·6 g dm^{-3} and acid content 1·4 g dm^{-3}. This permeate was almost colourless and had a less currant-like smell.

Double-blind experiments could not distinguish between the fresh juice and the semi-concentrate rediluted to the original solids content. The panelists rated the smell of the vacuum-evaporated sample significantly inferior, while the colour and taste of the two products were identical.

The ascorbic acid (AA) and dehydro ascorbic acid (DAA) quantity decreased during both reverse osmosis and vacuum evaporation. As expected the losses due to vacuum evaporation were larger. The AA:DAA ratio of the vacuum-evaporated product was less favourable (0·384) than the ratio obtained for the product prepared by reverse osmosis (3·53). This indicates that the more stable AA was retained to a larger extent, i.e. vitamin C level changes during storage will probably be more favourable.

The permeability of the membrane decreased significantly during the concentration step (Fig. 3).

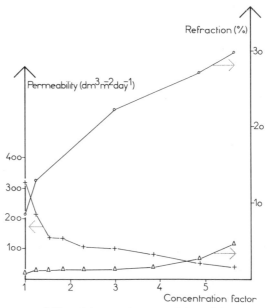

FIG. 3.   Water permeability of the membrane and the solids content of concentrate and permeate as a function of the concentration factor of red currant juice. Membrane = 80/6; pressure = 6·1 MPa; ○ = solids content of the concentrate; + = water permeability; △ = solids content of the permeate.

2.5.4. Blackcurrant Juice

A 2·4-fold concentration could be achieved for the initially 14% refractory level blackcurrant juice yielding a semi-concentrate of 30% solids content (Table 5). The permeate contained almost no colour material, even though concentration of this material in the starting solution was high. The acid and carbohydrate losses were also favourable. However, the permeability calculated for the whole process was low, 59 dm^3 m^{-2} d^{-1}, and even at the beginning of the process it remained below 100 dm^3 m^{-2} d^{-1} for a juice of 14% refractory content.

TABLE 5

CONCENTRATION DATA OF BLACKCURRANT JUICE SEMI-CONCENTRATE

Constituents and properties examined	Juice	Concentrate	Permeate	Losses %
Solids content (refractory %)	14·0	30·0	0·7	2·7
pH value	3·1	3·1	3·4	—
Density (kg.dm^{-3})	1·059	1·133	No data	
Acid content (calculated as citric acid g.dm^{-3})	27·9	62·0	2·3	4·8
Reducing sugars (calculated as glucose g.dm^{-3})	86·8	198.7	4·0	2·6

Pressure applied 6·0 MPa, pressure difference $\Delta P = 4·0 \rightarrow 0·3$ MPa (ΔP = pressure applied − approximative osmotic pressure); permeability average 59 dm^3/m^{-2}/dm^{-1} (98 → 15); rate of concentration 2·4; membrane 73/29; losses (%) calculated on the basis of the permeate.

The colour and smell of the fresh and rediluted juices could not be distinguished. However, the taste of the concentrate was found significantly better. This could be attributed to the fact that the unpleasant aroma components of blackcurrants were also lost in part during the membrane separation process.

2.5.5. Strawberry Juice

The data in Table 6 relate to a two-stage concentration process. A less selective membrane (73/14) was used in the first stage to produce a 26·7% concentrate from the initial 7% refractory content strawberry juice. The concentration factor was 5·3. It had a permeate of 1·9% solids content. This permeate was further concentrated by a 78/19 membrane and in a 16·6-fold concentration step the solid content became 20·5%. The two concentrates

TABLE 6
CONCENTRATION DATA OF STRAWBERRY SEMI-CONCENTRATE PRODUCED IN TWO STEPS

Constituents and properties examined	Juice	First step			Second step			Concentrate III (I + II)	Losses %	Concentrate (diluted to the volume of the feed juice)			
					Concentration of the permeate I					I		III	
		Concentrate I	Permeate I	Losses %	Concentrate II	Permeate II	Losses %			Absolute values	Differences	Absolute values	Differences
Solids content (refractory %)	7·0	26·7	1·9	21·6	20·5	0·6	29·5	25·4	6·4	5·5	−1·5	6·6	−0·4
pH value	3·5	3·3	3·6		3·4	3·5		3·3		3·5	0	3·5	0
Density at 17°C (kg.dm⁻³)	1·028	1·118	1·001		1·090	1·001		1·112		1·020	−0·008	1·025	−0·003
Acid content (calculated as citric acid) g.dm⁻³	9·1	33·9	2·8	25·0	31·5	1·5	48·7	33·4	12·2	6·4	−2·7	8·0	−1·1
Reducing sugars (calculated as glucose) g.dm⁻³	47·1	119·3	10·4	17·9	142·3	1·29	11·7	187·6	2·1	37·7	−9·4	44·7	−2·4

First step: Pressure applied $4\cdot0 \to 6\cdot0$ MPa; pressure difference $\Delta P = 3\cdot2 \to 1\cdot2$ MPa ($\Delta P =$ pressure applied − approximative osmotic pressure); permeability average $317 \, dm^3 \, m^{-2} \, d^{-1}$ ($576 \to 115$); rate of concentration $5\cdot29$; membrane 73/14; losses (%) calculated on the basis of the permeate I.

Second step: Pressure applied $6\cdot0$ MPa; pressure difference $\Delta P = 5\cdot9 \to 2\cdot8$ MPa; permeability average $708 \, dm^3 \, m^{-2} \, d^{-1}$ ($1242 \to 236$); rate of concentration $16\cdot59$; membrane 78/19; losses (%) calculated on the basis of the permeate II and referred to the losses of the first step.

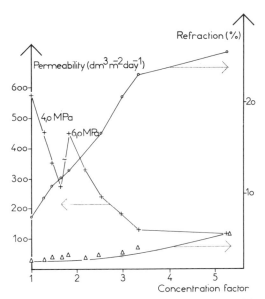

FIG. 4. Water permeability of the membrane and the solids content of the concentrate and the permeate as a function of the concentration factor of strawberry juice. Membrane = 73/27; pressure = 4·0–6·0 MPa; + = water permeability; ○ = solids content of the concentrate; ∧ = solids content of the permeate.

were mixed yielding a product with 25·4 % refractory content. The average permeability of the preconcentrating membrane was 317 dm³ m⁻² d⁻¹ (see Fig. 4), i.e. its value was several times higher than usual in single-step fruit juice concentration experiments. However, the permeate was pink and contained large amounts of carbohydrates and acids. The average permeability measured for the second concentration step of the permeate was 708 dm³ m⁻² d⁻¹ and the membrane completely retained the colour components. It could reduce the carbohydrate and acid losses of the first stage also.

Double-blind comparisons were carried out between the fresh juice and the rediluted first semi-concentrate; the fresh juice and the rediluted, unified semi-concentrate; and the rediluted first and final semi-concentrates. The panelists could detect no significant difference.

2.5.6. Apple Juice

The apple variety selected for the investigation, Jonathan, was extremely rich in low molecular weight, volatile aroma components. Therefore, it was

TABLE 7
CONCENTRATION DATA OF APPLE JUICE SEMI-CONCENTRATES PRODUCED BY REVERSE OSMOSIS AND BY VACUUM EVAPORATION

Constituents and properties examined	Juice	Reverse osmosis			Evaporation	Concentration (concentrates diluted to the refractory % of juice)			
						Reverse osmosis		Evaporation	
		Concentrate	Permeate	Losses %	Concentrate	Absolute values	Differences	Absolute values	Differences
Solids content (refractory %)	11·5	31·6	1·0	5·7	39·6	11·5	0	11·5	0
pH value	3·4	3·3	3·3	—	3·4	No data		No data	
Density at 17°C (kg.dm^{-3})	1·043	1·140	1·001	—	1·181	No data		No data	
Acid content (calculated as citric acid g.dm^{-3})	6·9	13·8	1·9	18·1	21·7	4·6	−2·3	5·6	−1·3
Reducing sugars (calculated as glucose g.dm^{-3})	76·0	229·4	4·4	3·8	285·1	76·4	+0·4	73·1	−2·9

Pressure applied 6·0 MPa; pressure difference $\Delta P = 4\cdot4 \rightarrow 0\cdot2$ MPa ($\Delta P =$ pressure applied − approximative osmotic pressure); permeability average 78 dm^3 m^{-2} d^{-1} (132 → 10); rate of concentration 3·2; membrane 73/29; losses (%) calculated on the basis of the permeate.

TABLE 8
CONCENTRATION DATA OF VEGETABLE JUICE SEMI-CONCENTRATES PRODUCED BY REVERSE OSMOSIS

Constituents and properties examined	Red beet			Carrot			Tomato		
	Juice	Concentrate	Losses %	Juice	Concentrate	Losses %	Juice	Concentrate	Losses %
Solids content (refractory %)	4·7	23·5	2·9	6·2	14·3	6·2	5·4	29·0	6·7
Acid content (calculated as citric acid g.dm^{-3})	3·5	17·1	8·7	No data		—	3·5	20·4	4·1
Reducing sugars (calculated as glucose g.dm^{-3})	1·5	8·2	1·8	2·3	4·9	4·3	36·2	203·9	6·3

Red beet: Pressure applied 5·8 MPa; permeability average 421 dm^3 m^{-2} d^{-1} (656 → 240); rate of concentration 5·7; membrane 77/9.

Carrot: Pressure applied 4·0 MPa; permeability average 120 dm^3 m^{-2} d^{-1} (171 → 68); rate of concentration 2·5; membrane 73/28.

Tomato: Pressure applied 6·0 MPa; permeability average 372 dm^3 m^{-2} d^{-1} (414 → 240); rate of concentration 6·0; membrane 77/9.

Losses (%): Calculated on the basis of the permeate.

expected, and found, that the characteristic taste and smell-influencing compounds entered the permeate. The results are shown in Table 7. The rediluted, 31·6% solids-containing semi-concentrate proved significantly inferior to both the fresh juice and the concentrate obtained by vacuum evaporation. However, the losses of reverse osmosis concentration were similar to those obtained in other single-step fruit juice concentration experiments, and did not substantiate the unfavourable opinion of the panelists.

2.6. Vegetable Semi-concentrates

The data shown in Table 8 relating to reverse osmotic concentration of vegetable juices was obtained from preliminary experiments. The juices were obtained the same way as the fruit juices, except that centrifugation was used for the separation of tomato serum.

The raw juices had a lower solids, acid and carbohydrate content than the fruit juices. The only exception was the tomato serum as its carbohydrate concentration was similar to that of the fruit juices. The permeation values at 6·0 MPa pressure were favourable for red beet juice and tomato serum. The membrane 77/9 was selective for the colour components and retained the carbohydrates and acids to an acceptable level. At a pressure of 4·0 MPa a 73/28 membrane could produce only a 14·3% juice from the 6·2% refractory raw juice.

3. GENERAL CONCLUSIONS OF THE CONCENTRATION EXPERIMENTS

Both the data reported in the literature and our experiences discussed above prove unambiguously that fruit and vegetable juices can be concentrated advantageously only to a certain extent by the existing cellulose acetate membranes and reverse osmosis apparatus.

With increasing solids content, and due to local high concentration spots on the surface of the membrane, the osmotic pressure becomes so high that at least 10·0 MPa pressure is required to produce concentrates of 40–45% solids content at acceptable rates. The technical requirements of large-scale units producing concentrates at the usual 65% or higher solids levels would make capital investment and operation costs unreasonably high. Also, new membrane materials suitable for long-term, high pressure operation without collapsing would have to be developed.

The quality of semi-concentrates produced by reverse osmosis is generally better than that of those produced by evaporation. Less aromatic, highly acidic fruits are more suitable for reverse osmosis than those which contain large amounts of low molecular weight, water-soluble aroma components. The quality of the concentrates could probably be further improved by the development of specially designed membranes and apparatus and also by optimising the concentration technologies.

The concentration losses are more favourable at low concentrations. With increasing solids content the amount of valuable components entering the permeate increases. The carbohydrate loss cannot be considered significant and it does not degrade the quality of the product. Partial loss of the acids can be an asset in certain cases, e.g. for the improvement of the carbohydrate–acid ratio of grape juice or sour fruit juices such as sour cherry and currant juices. The loss of acids is a liability in the case of low-acid fruits, where the acid level has to be restored for full taste. Occasionally additional acidification might be necessary if the product is to be stored. Our experiments proved that changes of the amino nitrogen content were higher and those of the polyphenols were lower in the semi-concentrates as determined from the permeates. Neither alcohols, nor hydroxy methyl furfurol were formed in the concentration step.

The partial removal of the aroma components is the major quality influencing concentration loss. Experience has shown that these low molecular weight, water-soluble components enter the first permeate. They cannot be fully retained with the known membranes, but there are several methods for their partial recovery and the reduction of the other losses as well. It has been proved experimentally that the two-stage concentration of fruit juices is more advantageous than the single-step process, because the concentration factor obtained is higher, the quality of the product is better, and the yield of the separation process is higher. The permeate, and especially its aroma-rich fractions can be recirculated into the concentrator either in continuous or batch operation mode. Since the membranes are saturated after a while, further valuable compounds can be recovered from the recirculated permeate. The efficiency and yield of the concentration process and the quality of the product can be improved by combining membranes of different characteristics. The selection of membranes selective for certain aroma components of various fruits is aided by theoretical statements and experimental results obtained with model solutions.[25–28] A not yet fulfilled condition of membrane selection is the knowledge of the characteristic aroma components of various fruits and their variants. Though the aroma components occuring in fruits are known, the selection of the most characteristic ones by sensory evaluation is not yet accomplished, so only indirect methods, e.g. gas chromatography, are available for the selection of the membranes displaying optimum selectivity for certain aroma components.

The approximate ΔP values, calculated according to ref. 14 and shown in the tables, explain why the average permeabilities are so low. For the concentration process the average driving force is only a few tenths of a MPa. Therefore, 6·0 MPa pressure is insufficient for the preparation of concentrates with 30% solids content in a single-step concentration process. At least 8·0 MPa is required to ensure a maximum of 2·0 MPa pressure difference even at the end of the separation process. Fruit and vegetable juices can be concentrated advantageously in a two-stage process in which the higher molecular weight components are concentrated at low pressure by a more 'open' membrane in the preconcentration step, and the permeate of this step containing low amounts of solids is concentrated, also at low pressure, by another, more selective membrane of lower permeability. This arrangement significantly boosts the productivity of membranes and decreases the residence time of juices.

4. STORAGE AND THE USE OF SEMI-CONCENTRATES

Generally, fruit juices are stored as concentrates containing 65% solids or as superconcentrates of even higher solids content. These products are microbiologically safe. However, investigations are under way to determine the possibilities of longer storage without the deterioration of the quality of the product. The major goals of this research activity are the discovery and control of chemical reactions influencing the colour and taste-determining aroma components.[29,30]

Due to their high water content the semi-concentrates are less stable than the concentrates and the conditions of their storage are not yet fully known. Their industrial utilisation is hardly known, so they are to be considered as new products.

Experiences gained in the course of storage experiments indicate that the highly acidic sour cherry and currant semi-concentrates prepared by reverse osmosis can be stored at room temperature over six months if $0·6–1·0 \, g \, dm^{-3}$ potassium sorbate is added as preservative. Their taste and smell is better than that of the control sample stored under identical conditions and prepared by vacuum evaporation. Under identical conditions peach and apple juice concentrates became unsuitable for consumption after only three months. Stored at $+3 \, °C$ sour cherry, red currant and blackcurrant semi-concentrates preserved their taste, colour and aroma for at least 12 months.

Cooling increases storage costs. Yet this is one of the future directions even in the case of highly concentrated fruit juices, because cooling can preserve their quality for longer periods.[29] Further research is needed for the development of the optimum preservation and storage methods of semi-concentrates, this is one of the prerequisites of their industrial utilisation and wide-spread consumption.

There are a number of possibilities for the use of fruit and vegetable semi-concentrates in the food industry.

Since part of the acid content of the juices is lost into the permeate the carbohydrate–acid ratio of the product changes. Therefore, dietetic soft drinks can be produced from semi-concentrates without, or only with moderate addition of sweeteners.

After sugar addition and perhaps restoration of the acid level to the original value the semi-concentrates can be used to produce regular soft drinks with increased juice proportions.

Since dilute solutions can be processed economically by reverse osmosis, its combination with other concentration methods seems advantageous because its energy consumption is low.

Powders produced from semi-concentrates might become high-quality colour and taste improving additives in ice cream, creams, cake-covers, sweet or sour dairy products.

5.   ENERGY CONSUMPTION OF THE PROCESS

The major characteristics of fruit juice concentrates containing 30–35% solids and prepared by reverse osmosis are superior to those obtained by thermal treatment. However, without a firm, market-proved advantage the more favourable characteristics of fruit juices prepared by reverse osmosis cannot be considered when the economics are calculated.

In economic forecasts concerning industrial-scale introduction of fruit juice concentration by reverse osmosis, the advantages of the new process manifest themselves first in energy savings and second in capital costs.

A short review, based partly on the literature[14,31,32] as well as our data and calculations is presented here comparing the expected economics of reverse osmosis and the proven, known values of two-stage thermal evaporation.

Since thermal evaporation requires heat energy while reverse osmosis requires electric energy, a comparison factor of 3 was accepted which also accounts for the inherent inefficiencies and losses.

The energy consumption of the concentration process depends on the composition of the solution. Our data refer to a concentration process which produces a juice of 50% solids content from fresh juices of 10% refractometric solids content. Since, on average, maxima of 35% final concentration could be achieved by reverse osmosis, this ensures a considerable safety margin to account for unexpected energy losses. It can be calculated from the free energies of the initial and final solutions and the behaviour of real solutions that a maximum of 4 kJ are required for the removal at 20 °C of 1 kg water from solutions within the above concentration limits. This would require a heat energy input of 12 kJ. This value is so low that it has never been realised in any practical thermal concentration process.

In the case of simple reverse osmosis (without recirculations and energy recovery), thermodynamic analysis predicts a minimum of 19·0 kJ energy requirement for the removal of 1 kg water. This corresponds to 57 kJ heat energy introduced by steam.

The selected 50% final solids content level requires 16 MPa osmotic pressure. This is a hypothetical value because at present both the apparatus and the membranes readily available are limited to the 4–9 MPa range. This pressure corresponds to a final solids content of 25–45 %, depending on the dissolved components.

In practice, however, the amount of energy required for the removal of 1 kg water by reverse osmosis is considerably higher than the above theoretical minimum value. For practical permeation rates, operation pressures considerably higher than the osmotic pressure have to be applied, because of the polarisation phenomena taking place on the surface of the membrane itself. Also, the efficiency of the pumps used today is rather low, 10–80 %, depending on their construction.

Multistage centrifugal pumps can be used to 2·5 MPa but their efficiency is very low. The efficiency of expensive piston pumps providing pressures as high as 15 MPa is higher (75–80 %).

On average the electric energy required for the removal of 1 kg water is 160 kJ (equivalent to 480 kJ heat energy delivered by steam) in the case of low pressure, low efficiency pumps. In the case of high efficiency, high pressure pumps this figure is a more favourable one of 25 kJ (75 kJ heat energy). However, this value does not contain the energy required to counter the effects of polarisation, membrane resistance and pressure. The energy consumption of the latter phenomena can be decreased by moderating polarisation (using recirculation by a high pressure recirculating pump) and increasing the permeability of the membrane while maintaining its selectivity.

Goldsmith and Horton[33] reported that 28·5 kJ electrical energy (equivalent to 85·5 kJ heat energy) was required for the removal of 1 kg water in a circulating reverse osmosis apparatus using pumps with efficiencies amounting to 70% and membranes with permeation rates as high as 325 dm^3/m^{-2} d^{-1} for dilute and 80 dm^3 m^{-2} d^{-1} for concentrated solutions.

This value agrees with the figure calculated for systems equipped with high efficiency, high pressure pumps. It also indicates that in the case of suitable membranes and apparatus the energy required to counter the effects of polarisation, membrane resistance and pressure losses can be as low as 15% of the calculated minimum energy requirement of the operation.

An average dual-stage thermal evaporator equipped with an aroma recovery unit and used for the concentration of liquid food products requires 690 kJ heat energy in the form of steam for the removal of 1 kg water. This figure is nine times higher than the energy requirement of reverse osmosis.

A number of authors compared the costs of various water removal processes.[31−37] The publication of Bomben[31] deserves special attention. According to his analysis only small reverse osmosis apparatus (max. 500 kg h^{-1} water removal rate) are competitive with the dual-stage thermal evaporators equipped with aroma recovery units. His data indicate that the capital costs of a reverse osmosis apparatus removing 1000 kg water in an hour are twice as high as those of a similar capacity, dual-stage thermal evaporator equipped with an aroma recovery unit. A permanent cost factor of reverse osmosis apparatus is the periodic replacement costs of the membrane itself. If the membrane is replaced only once a year then its costs amount to 10–20% of the permanent costs. No doubt, the capital costs of reverse osmotic apparatus decrease much less with increasing unit size than do those of a thermal evaporator.

However, the present energy crisis and higher energy costs favour reverse osmosis. In comparison with the 1971 data reverse osmosis units as large as 2000 kg h^{-1} are competitive with the equivalent thermal evaporators.

Detailed economic analysis of the operation data of existing reverse osmosis apparatuses could provide a reliable basis for the wide-spread industrial application of this new technique.

REFERENCES

1. WOOLLEN, A., *Fd Mf.*, 1978, **53**, 33.
2. DU PONT DE NEMOURS AND CO., Prospectus, Wilmington.
3. HORTON, B. S., GOLDSMITH, R. L. and ZALL, R. R., *Fd Technol.*, 1972, **26**, 30.
4. NIELSEN, I. K., BUNDGARD, A. G., OLSEN, O. J. and MADSEN, R. F., *Process Biochem.*, 1972, **7**, 17.
5. MADSEN, R. and OLSEN, O. J., *Chemie Techn.*, 1974, **3**, 81.
6. RÖDICKER, H. and KROLL, U., *Lebensmittel Ind.*, 1976, **23**, 303.
7. KARDOS, E., *Flüssiges Obst*, 1980, **47**, 190.
8. MORGAN JR., A. I., LOWE, E., MERSON, R. L. and DURKEE, E. L., *Fd Technol.*, 1965, **19**, 1790.
9. MERSON, R. L. and MORGAN JR., A. I., *Fd Technol.*, 1968, **22**, 97.
10. GHERARDI, S., PORRETTA, A. and DALL'AGLIO, G., *Industria Conserve*, 1972, **47**, 16.
11. NOMURA, D. and HAYAKAWA, I., *Nippon Shokuhin Kogyo Gakkaishi*, 1976, **23**, 404.
12. WATANABE, A., KIMURA, S., UMEDA, K. and KIMURA, S., 'IUFOST 79' Symp., 1979, Helsinki.
13. BOLIN, H. R. and SALUNKHE, D. K., *J. Fd Sci.*, 1970, **5**, 211.
14. THIJSSEN, H. A., *J. Fd Technol.*, 1970, **5**, 211.
15. POMPEI, C. and RHO, G., *Lebensmittel Wiss. u. Technol.*, 1974, **7**, 167.
16. PEYNAUD, E. and ALLARD, J., *Cptes Rendus Acad. de France*, 1970, **56**, 1454.
17. PERI, C. and POMPEI, C., *Vini d'Italia*, 1975, **17**, 179.
18. SCHOBINGER, U., *Schweiz. Z. Obst u. Weinbau*, 1972, **108**, 572.
19. WUCHERPFENNIG, K., *Anwendungsmöglichkeiten von Membranprocessen bei der Herstellung von Getränken*, 1st edn, 1977, Offset Köhler, Giessen.
20. WUCHERPFENNIG, K. and NEUBERT, S., *Flüssiges Obst*, 1977, **44**, 15.
21. WUCHERPFENNIG, K. and NEUBERT, S., *Flüssiges Obst*, 1977, **44**, 46.
22. RHEIN, O., *Die Weinwirtschaft*, 1975, **111**, 134.
23. SCHOBINGER, U., KARWOWSKA, K. and GRAB, W., *Lebensmittel Wiss. u. Technol.*, 1974, **7**, 29.
24. DRIOLI, E., ORLANDO, G., D'AMBRA, S. and AMATI, A., 'IUFOST 79' Symp., 1979, Helsinki.
25. MATSUURA, T., BAXTER, A. G. and SOURIRAJAN, S., *J. Fd Sci.*, 1975, **40**, 1039.
26. MATSUURA, T., BAXTER, A. G. and SOURIRAJAN, S., *Acta Aliment. Acad. Sci. Hung.*, 1973, **2**, 109.
27. PERI, C., BATTISTI, P. and SETTI, D., *Lebensmittel Wiss. u. Technol.*, 1973, **6**, 127.
28. MATSUURA, T. and SOURIRAJAN, S., *AICHE Symp. Ser.*, 1978, **74**, 196.
29. GHERARDI, S., 'IFU 80' Symp., 1980, Bled.
30. KLÄUI, H., 'IFU 80' Symp., 1980, Bled.
31. BOMBEN, J. L., Aroma recovery and retention in concentrating and drying of foods. In *Advances in Food Research*, Vol. 20, 1st edn. (Chichester, C. O. *et al.*, Eds.), Academic Press, New York, 1973.
32. SCHWARTZBERG, H. G., *Fd Technol.*, 1977, **31**, 67.
33. GOLDSMITH, R. L. and HORTON, B. S., *Wat. Pollut. Control Res. Ser.*, 1971, 12060 DXF.

34. WYSOCKI, G., *Chemie Technik*, 1976, **5**, 177.
35. VAN PELT, W. H., 'Separation Processes by Membranes, Ion Exchange and Freeze-concentration in the Food Industry' Symp., 1975, Paris.
36. MARQUARDT, K., *Chemie Technik*, 1975, **4**, 289.
37. PERI, C., *Sci. Technol. Alimenti*, 1974, **4**, 43.

Chapter 4

THE EFFECT OF MICROWAVE PROCESSING ON THE CHEMICAL, PHYSICAL AND ORGANOLEPTIC PROPERTIES OF SOME FOODS

MARGARET A. HILL
Department of Food Science and Nutrition,
Queen Elizabeth College, University of London,
London, UK

SUMMARY

Investigations have been carried out regarding the feasibility of the application of microwave energy for enzyme inactivation (including blanching), dehydration, freeze drying, pasteurisation, sterilisation and defrosting of frozen foods. These processes and their effect on the chemical, physical and organoleptic properties of some foods are reviewed.

Cooking by microwaves is now relatively well established and the flavour of a number of foods heated by microwave energy is discussed. For reference purposes some published research on the dielectric properties of food materials during the past 10 years is listed.

1. INTRODUCTION

The literature is continually expanding regarding microwave applications to food. Energy savings and a reduction in processing time are two advantages of microwave heating which are pertinent at the present moment.

The interaction of food and microwave energy is complex. To assist understanding and enable predictions of near optimum methods to be made further knowledge of the dielectric properties of foods, actual

121

temperature profiles within foods during cooking or processing and other data which determine food quality, e.g. sensory parameters, moisture contents, etc., is required. Computers are beginning to be used to assist in these studies, and in the future, more optimisation studies on the application of microwaves to food should occur.

2. DIELECTRIC PROPERTIES OF FOOD

In 1971 Bengtsson and Risman[1] stated that 'Knowledge of the dielectric properties of food materials is important for proper understanding of the heating pattern during microwave heating of foods, such as in the cooking of foods or in reheating of precooked foods from frozen or refrigerated condition. Unfortunately such data are missing to a considerable extent, in spite of a few noteworthy publications over the last decade'.

In the same paper Bengtsson and Risman presented measurements of the dielectric properties of a number of foods at 3 GHz using the cavity perturbation technique.[2] In subsequent years the following work appeared:

1972—Roebuck and co-workers[3] investigated the dielectric properties of carbohydrate–water mixtures at microwave frequencies of 1 and 3 GHz over a wide compositional range.

1974—Tinga and Nelson[4] prepared a comprehensive table of the dielectric properties of materials for microwave processing.

1974—To and co-workers,[5] in an attempt to establish a predictive model and a more basic understanding of dielectric properties, made measurements (using the precision slotted line technique) on rehydrated non-fat dry milk at temperatures of 25, 35, 45 and 55 °C and at frequencies of 300, 1000 and 3000 MHz. The dielectric loss factors did not agree with the values predicted by chemical composition; this was in accord with data presented by Mudgett et al.[6] in 1971. The measured values were lower due to a complex relationship of solute–solute and solute–solvent interactions.

1975—Ohlsson and Bengtsson[7] using a cavity perturbation method presented dielectric data for microwave sterilisation processing.

1976—Metaxas[8] reviewed the properties of materials at microwave frequencies and concluded that the first step in considering whether it is advantageous to treat a certain product with microwave energy is to determine the dielectric properties at the frequency concerned. Conditions must however be standardised if the data are to be meaningful.

It was also stressed by Bengtsson and Risman[1] that although knowledge of the dielectric properties of foods and the conditions which influence them is fundamental to enable predictions of dielectric heating of foods to be made, they do not completely explain the heating patterns of food in a microwave field. Other factors which influence the heating of food in a microwave field are applicator or oven design, distribution and geometry of the load,[9] standing wave patterns in the load, the degree of heat and mass exchange with the surroundings, and also the thermal properties of the food, such as density enthalpy curve and thermal conductivity.

1980—Mudgett et al.[10] published work on the dielectric behaviour of a semi-solid food at low, intermediate and high moisture contents.

No attempt has been made here by the author to review the work presented in these preceding papers but it is suggested that prior to embarking on experiments with microwaves and food the relevant dielectric data are consulted.

3. ISM FREQUENCIES

An organisation known as the International Telecommunication Union (ITU) has been established by the nations of the world. The ITU aims at establishing regulations which permit the orderly use of the radio frequency spectrum. A set of frequencies within the frequency spectrum have been established which are set aside for use with industrial, scientific and medical (ISM) apparatus.

The two frequencies used for cooking and food processing are 2450 MHz $\pm$ 50 MHz and 915 MHz $\pm$ 25 MHz.[11,12]

4. INDUSTRIAL APPLICATIONS

4.1. Enzyme Inactivation

All the evidence indicates that the inactivation of enzymes by microwaves is due to a thermal effect.[13-15] Conventional heat sources are frequently used in food processing to inactivate enzymes. Investigations on a number of foods have been carried out to determine whether microwave energy can be used to replace conventional methods.

4.1.1. Fruit Juice

Copson[16] described a method and apparatus for inactivating enzymes in fruit juice. In particular pectin-methylesterase which is present in raw fruit juice should be inactivated. Pectin-methylesterase catalyses the hydrolytic removal of methoxyl groups from the pectin molecule to form low ester pectins which results in the separation of the liquids and solids in the juice, i.e. the cloudy appearance of the fruit juice is lost. Microwave energy was applied at 2450 MHz so that fruit juice reached temperatures in the range of 60–80 °C. Flavour tests on orange juice concentrate processed in this way showed that the natural flavour and quality of the juice was not impaired.

4.1.2. Wheat Flour

Excessive α-amylase activity in flour is undesirable[17] because it results in the production of excess dextrins during the fermentation of the dough which in turn cause bread to be produced with a tough, soggy crumb. Flours which are milled from sprouted grains are high in α-amylase activity and this can occur in Canadian wheats when wet weather occurs just prior to harvest. It is therefore often necessary to inactivate α-amylase and the use of microwaves for this purpose has been investigated.[18] A microwave cooker operating at 2450 MHz with a total output of 1·8 kW was used. A 60 s exposure of the flour (500 g) reduced the enzyme activity to an acceptable level without any deleterious effects on the principal characteristics of the flour, related to dough formation. The number of viable organisms in the flour were reduced considerably but there was a high moisture loss.

4.1.3. Peaches

Avisse and Varoquaux[19] explain that the browning of peaches which occurs rapidly during freezing, frozen storage and thawing can be reduced by:

1. Selecting a peach variety with a low polyphenol content
2. Limiting the quantity of oxygen available for the oxidation of phenolic compounds, e.g. by covering the fruit with syrup containing sulphur dioxide or ascorbic acid prior to freezing
3. Inactivating the enzymes peroxidase and polyphenoloxidase by blanching

These authors blanched peaches in a conventional domestic microwave oven operating at a frequency of 2450 MHz with a power output of 700 W. The samples were heated with 1 min impulses, 1 min on and 1 min off, the time of exposure being 0 (control), 6, 8 or 10 min.

The peaches were then cut in half, the stones removed and immediately frozen in liquid nitrogen. The process was evaluated regarding temperature, enzyme inactivation of peroxidase and polyphenoloxidase, colour, flavour and liquid exuded after thawing.

Ten minutes microwave exposure was found necessary to inactivate nearly completely the enzymes peroxidase and polyphenoloxidase in the four varieties tested. When these samples were tasted they did not taste as if they had been cooked in boiling water. All the samples, however, showed brown spots under the skin, this could be due to the higher polyphenol and polyphenoloxidase content of the skin or because the microwave heating did not inactivate enzymes on the surface.

Table 1 shows the temperature of whole peaches after microwave heating.

It would therefore seem necessary to either remove the skin or use a combination method of blanching, e.g. hot water or steam and microwave.

TABLE 1

TEMPERATURES IN °C OF PEACHES AFTER MICROWAVE HEATING[a]

	Microwave heating time (min)		
	6	8	10
Inner part of peach	65	75	85
Under the peel of the peach	40	45	50

[a] Adapted from Avisse and Varoquaux.[19]

4.1.4. Potatoes

Browning occurs if potatoes are not blanched prior to freezing. Polyphenoloxidase is credited to be chiefly responsible for the rapid browning of cut potato surfaces although peroxidase may also be involved.[20,21] Conventional blanching, i.e. using hot water or steam to inactivate enzymes, often results in some advance effect on the product, e.g. an excess of water soluble vitamins may be lost and the product may be overcooked. Blanching by microwave energy has therefore been investigated.[22,23]

Collins and McCarty[22] compared two methods of blanching potatoes, the potatoes were immersed in boiling water prior to further heating by microwaves or by conventional means. A time of 13 min was required to inactivate the enzyme peroxidase to the core of the potatoes which had a near radius of 22·7 mm when conventional heating was used. A time of only 4·7 min was necessary when potatoes of the same size were exposed to microwave energy. A texture difference was apparent, the microwave blanched potatoes were softer.

The possibility of blanching potatoes with 'dry' microwave heating followed by immersion in boiling water has been investigated.[23] Small whole white potatoes (mean weight 29 g) were heated in a microwave cooker for 0·5, 1·0, 1·5 and 2·0 min followed by 0, 1, 2, 3, 4 and 5 min immersion in boiling water. The presence of two heat gradients was established, one from the core and one from the periphery. The heating times required to inactivate peroxidase completely across the mean radius were 1·5 min microwave followed by 3 min boiling water or 2 min microwave and 2 min boiling water. It was reported that heating for 1 min or less in the microwave cooker with subsequent treatment with boiling water for up to 5 min did not inactivate peroxidase completely. These results showed that microwave energy together with hot water treatment could be used to shorten the blanching time and to obtain a more uniform texture throughout the potato.

Other workers[24,25] have used this two-stage process to blanch standard cylinders of potato (diameter 25 mm and length 50 mm cut from the long axis of potato tubers). Five potato cylinders required 3-min treatment with microwave (2450 MHz) followed by 0·25 min in boiling water, however the variation in chemical composition of potatoes made it imperative for new process times to be determined for each batch of potatoes. These findings are in agreement with those of Chen et al.,[23] that microwave energy followed by immersion in boiling water shortens the time required for blanching when compared to blanching in boiling water only.

With the aim of relating blanching time to moisture content Abara and Hill[25] determined the blanching times of standard cylinders of potato from four varieties using microwave energy followed by immersion in boiling water. No conclusive results were obtained regarding moisture content and blanching time but a relationship was shown between peroxidase activity and blanching time. The greater the peroxidase activity per gramme of potato the longer the blanching time.

4.1.5. Corn-on-the-cob

A continuous microwave tunnel operating at 915 MHz was used by Huxsoll et al.[26] for blanching corn-on-the-cob. They also used a combined method of water and microwaves. A 4-min water blanch followed by a 2-min exposure to microwave energy was equivalent to a 6-min blanching time using microwaves only. Slight peroxidase activity remained in the cob. For cost reasons they advised substitution of conventional energy as much as possible, however for the combined method when they doubled the water blanch time and reduced the microwave blanch time by one half the enzyme inactivation was insufficient.

Dietrich *et al.*[27] compared steam blanching with microwave heat alone and a combination method of hot water and microwaves. The cooked products after freezing were ranked in order of best natural corn flavour or least off-flavour, with the best samples receiving the lowest ranks. None of the samples which were blanched by microwave heat alone, or a combination of hot water followed by microwave heating, were ranked better than the steam blanched sample either initially or after a year of storage at $-18\,°C$. Results of these experiments do not therefore indicate any new advantage in quality or stability that can be ascribed to microwave heating. A possible advantage mentioned is that steam injection into the microwave cooker for a simultaneous blanch probably would give a shorter blanch time and result in less area being occupied by equipment.

4.1.6. Broccoli

A number of studies on the blanching of broccoli for freezing have been reported. Table 2 shows the percentage retention of ascorbic acid by different blanching methods from the results of a number of workers.

As can be seen, microwave blanching[28] and steam blanching[29] gave the highest retention of ascorbic acid. From Noble and Gordon's experiments[29] the steam blanching method showed a greater retention of ascorbic acid 79 % (21 % loss) than water blanching 57 % retention (43 % loss). However, this difference disappeared when the samples were cooked, the average retention being approximately 29 % of the original content.

Eheart's[28] work included microwave blanching and in this instance the higher retention of ascorbic acid was maintained after frozen storage and

TABLE 2

RETENTION OF ASCORBIC ACID IN BROCCOLI DURING BLANCHING[a]

Method	Temp. (°C)	Time (min)	Ascorbic acid retention (%) wet basis	Reference
Water	100	4	57	Noble and Gordon[29]
	77	3	61 ⎫	Eheart[28]
		10	55 ⎭	
Steam		5	79	Noble and Gordon[29]
Microwave		2	79	Eheart[28]

[a] Reproduced with permission from *J. Am. diet. Ass.*, 1967, **50**, 207 and 1964, **44**, 120. Courtesy of The American Dietetic Association.

cooking when compared to the other methods reported. As well as ascorbic acid, Eheart also studied the effect of microwave blanching and water blanching on total acid, pH and the chlorophyll content of broccoli. Microwave blanched broccoli was higher than water blanched samples in total acids (wet basis) and percentage retention of chlorophyll *a*. After frozen storage and cooking microwave blanched samples were lower in pH, higher in total acids and in ascorbic acid (wet basis) but lower in total chlorophyll than broccoli blanched in 100 °C water. The higher chlorophyll losses in microwave blanched broccoli were thought to be due to the lower pH of these samples. However, prior to freezing the total chlorophyll retention and ascorbic acid content were highest for microwave blanching, as shown on Table 3. Retention of chlorophyll is of some importance because Sweeney *et al.*[30] showed that percentage retention of chlorophyll in cooked broccoli correlates directly to sensory panel scores for colour.

The results of these studies clearly indicate that when evaluating a blanching method more than one parameter should be measured to give an overall evaluation of the quality of the product.

TABLE 3

EFFECT OF THE BLANCHING METHOD ON THE ASCORBIC ACID AND CHLOROPHYLL CONTENT OF BROCCOLI PRIOR TO FREEZING[a]

Blanching method	Ascorbic acid wet basis (mg/100 g)	Retention of chlorophyll total (%)
Raw	100·2	95·2
100 °C water 3 min	60·8	86·4
77 °C water 10 min	55·6	74·1
Microwave 2 min	79·2	89·9

[a] Reproduced with permission from *J. Am. diet. Ass.*, 1967, **50**, 207. Courtesy of The American Dietetic Association.

4.1.7. Brussels Sprouts

A comparison of three blanching methods, microwave, conventional and a combination method was carried out by Dietrich *et al.*[31] After blanching the sprouts were frozen and the main aim was to achieve flavour stability without noticeable colour deterioration. For good colour retention a chlorophyll conversion to pheophytin of between 0–25 % is satisfactory. This can easily be exceeded because a long blanching time is often required for sprouts to achieve flavour stability. The minimum blanching time by each method was therefore found and the methods compared with respect to enzyme inactivation, heat penetration, chlorophyll, flavour and ascorbic acid retention. Storage stability at temperatures of $-29\,°C$, $-18\,°C$ and $-7\,°C$ was compared.

After a 6-min microwave blanch peroxidase in the centre of the sprouts was inactivated and no pink centres were seen but there was pink discoloration and peroxidase in the outer leaves. Excessive dehydration also occurred and off-flavours developed on storage. Steam blanching for 6–8 min at approximately $100\,°C$ was adequate for flavour stability both initially and after storage at $-18\,°C$ for six months and $-7\,°C$ for two months. A combination blanching method using microwave energy and steam or water to inactivate peroxidase produced sprouts which were flavour stable at $-29\,°C$, $-18\,°C$ and $-7\,°C$. Chlorophyll and ascorbic acid retention was as good or better than conventionally processed samples. It was concluded that concurrent application of steam and microwave heating for about 2 min may achieve an adequate blanch with only 15–20 % chlorophyll conversion.

A combined method using steam and microwave for blanching Brussels sprouts has been described by Meredith.[32] At the moment in the UK microwaves are not being used industrially for blanching, however a large-scale trial using a prototype 30 kW machine (Magnetronics Ltd, Leicester, UK) will soon be complete. For economic and technical reasons a combination method is recommended. A steam tunnel to raise the temperature of the product to $60\,°C$ followed by microwave heating is the most economic process.[33] The quality of the treated sprouts so far has been described as being as good as conventionally blanched sprouts and in some instances better, particularly when lower grade sprouts are used.

A patent[34] which contains details of equipment design and methods of blanching vegetables has been described. Four stages are recommended:

1. Preliminary surface blanch using hot water or steam at atmospheric pressure
2. Application of microwave energy (915 MHz)
3. Hot water blanch
4. Quenching in cold water

Details for the application of this procedure for blanching corn-on-the-cob, broccoli, cauliflower and asparagus are included in this patent.

4.2. Sterilisation of Solid Food

The aim of this process is to destroy food-spoilage and food-poisoning organisms and in particular the viable cells and spores of the bacteria *Clostridium botulinum*. The destruction is a time/temperature relationship; *Clostridium botulinum* spores are normally destroyed by 3 min at 121 °C but with conventional heating of food in cans it may take $1\frac{1}{2}$ h or more for the centre of the food to reach this temperature. Meanwhile, the surface will have been at this temperature for some time and the food will be overcooked and the organoleptic properties of the product may be impaired.

Figure 1[35] illustrates the time/temperature history for steam sterilisation of food in a can and in a pouch; it also shows microwave sterilisation of food in a pouch.

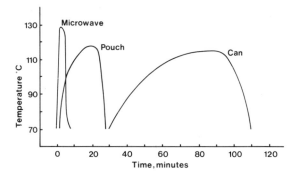

FIG. 1.   Time/temperature histories for steam sterilisation in a can, in a pouch and in microwave sterilisation (from Ohlsson[35]).

Heat is conducted slowly from the surface to the centre during the sterilisation of cans by the conventional method, hence the long processing time. However, by reducing the thickness of the sample, e.g. by sterilising the food in an aluminium foil pouch, very much faster heat transfer can be achieved and the quality of the food is improved as there is less thermal damage. As can be seen, an even faster rate of heating is obtained by microwave sterilisation. The thickness of the two pouches is identical.

Kenyon *et al.*[36] in 1971 selected microwave energy for sterilisation of meat in flexible pouches because very rapid heating times can be obtained. The process they developed involved passing food packaged in sealed plastic pouches through a microwave energy field on a continuous basis to achieve a sterilisation temperature. The pouches were then cooled and over packaged aseptically with a suitable plastic–foil laminate. This is necessary to protect the product on storage and during handling. Since temperatures in the order of 121 °C or greater are obtained within the pouch during microwave processing, internal pouch vapour pressure corresponding to temperature will exceed the rupture point of flexible pouches. An external air counter pressure (range 30–45 psig) is therefore provided to maintain pouch integrity during heating.

Further work done in the US Army Natick Laboratories[37] showed that the temperature spread between the centre and the surface during the sterilisation of beef slices was quite large. They used paper thermometers which were sealed in polypropylene envelopes, the resulting thermogram covered the rectangular slices of beef and were placed between the two 0·5-cm slices, inside each pouch. The work showed that the slowest heating point was at or near the geometric centre of the product where a temperature of 121 °C was reported, whereas at the outer edges the temperature recorded was 138 °C.

Similar results have been reported elsewhere. To achieve low temperature spread between the corner edges and the central parts, Stenström[38] and Ohlsson[35] stressed the necessity of immersing the food to be sterilised in a liquid during microwave heating.

Equipment has been designed[35] to carry out this process and an almost uniform temperature distribution has been achieved; for example a variation of ±2 or 3 °C just after microwave heating has been reported. A comparison of beans and peas sterilised by this process and by conventional canning, using a sensory panel, showed the microwave product to be superior in quality (see Table 4).

O'Meara et al.[39] agree that organoleptic evaluation of several foods demonstrated a quality advantage in favour of a microwave/hot water sterilising process rather than hot water heating alone. No data however to substantiate this statement were presented by these authors.

One problem which has mitigated against the adoption of microwave

TABLE 4
SENSORY COMPARISON BETWEEN CANNED AND MICROWAVE STERILISED PEAS AND BEANS (FROM OHLSSON[35])

Product	Process	Taste	Off-odour	Appearance	Texture
Peas	Canned	3·3	4·0	2·7	3·0
	Microwave	4·7	2·2	4·7	4·5
Beans	Canned	3·5	3·7	2·0	2·3
	Microwave	3·8	2·8	3·5	3·3

sterilisation is the availability of a suitable packaging material which will allow the passage of microwaves but prevent the absorption of oxygen and water. Hecht and Haskell[40] reported that a film consisting of a thin coating of an inorganic metallic phosphate, which is sandwiched between a polyester film and a heat-sealable layer, are compatible with microwave heating and have barrier properties necessary to provide shelf-stable packaging. A test run was carried out at the US Army Natick Laboratories; beef was packaged in a laminate containing this inorganic coating and then sterilised in a continuous microwave unit. The product showed no sign of spoilage after 10 months at room conditions.

They point out however that the barrier coating is lost if the inorganic-coated films are subject to retorting conditions. The sterilisation process at Natick was successful because the film was not exposed to water at high temperatures for too long.

Microwave sterilisation of oranges[41] in metal capped glass jars has been shown to be satisfactory on an experimental basis. Ascorbic acid retention was slightly better than the conventional method.

4.3. Sterilisation of Liquid Foodstuffs

An application of microwave heating for sterilising milk was described by Sale.[42] The whole process of heating, holding and cooling was carried out with a free jet of milk not touching the walls of the equipment. The sterile product produced was virtually indistinguishable from the starting material. However, although the process was technically successful, the process was too costly to develop. The same process has been shown to be technically feasible[43] for sterilising cream and orange juice. An expert taste panel were not able to detect flavour or other differences between treated (temperature attained 120 °C for 0·15 s) and untreated samples.

4.4. Pasteurisation

Experiments have been reported to determine the feasibility of pasteurising food by microwave energy.

4.4.1. Cured Hams

Cured hams[44] have been pasteurised by microwaves at both 60 MHz and 2450 MHz. The advantages were reduced juice loss and a shorter processing time in comparison with hot water processed hams. The serious disadvantage being that at both frequencies the surface temperature was lower than in the conventional hot water treatment, resulting in higher surface infection.

4.4.2. Milk

Milk[45] has been pasteurised by a batch and a continuous method. Bacteriological tests indicated that for the eight samples taken, the pasteurisation efficiency varied and this was attributed to the fact that it was difficult to maintain the processing milk temperatures at 82 °C throughout the run. Improved equipment should overcome this but as with the sterilisation of milk the high cost of equipment is a major drawback to the application of this technique on a large scale.

4.4.3. Soybean Curd

The shelf life of fresh soybean curds[46] has been extended by pretreating the packaged product with microwave energy at 2450 MHz. Comparative temperatures of 65, 80 and 95 °C were obtained in the product. Storage temperatures of 4·5, 13 and 21 °C were used. Control packages of soybean curds which were not treated by microwaves were also prepared. The flavour of the curds was evaluated periodically by using a sensory panel (an experienced panel of six tasters). Other measurements such as changes in pH, titratable acidity and viable microbial counts were made on the soaking water which was also packaged to protect the delicate curd.

On storage a decrease in pH, an increase in titratable acidity and the presence of viable bacteria in the soaking water accompanied decreased flavour scores of the soybean curds. Table 5 shows the flavour scores when the product was stored at 4·5 °C and demonstrates the extended shelf life which occurs when the microwave treatment was used.

The authors[46] indicated that this process was economically viable on an industrial scale and that a conveyor system and a microwave tunnel would be the most suitable method.

4.4.4. Oysters

A method for pasteurising oysters and similar products has been patented.[47] Freshly shucked oyster meat was fed through a microwave oven (range 2450–2500 MHz) for approximately $1\frac{1}{2}$ min, the temperature of the product was raised to about 60 °C and this was sufficient to pasteurise the oysters but maintain their firmness and taste when compared to freshly shucked oysters.

4.5. Dehydration

4.5.1. Pasta

Conventional drying of pasta takes 5–10 h and both temperature and humidity are carefully controlled so that the moisture is slowly driven out

TABLE 5

EFFECTS OF IN-PACKAGE MICROWAVE TREATMENT ON FLAVOUR
QUALITY OF STORED SOYBEAN CURDS[a]

Days of storage at 4·5°C	Flavour quality score (1 to 9)[b]			
	Control	Microwave treatment (°C)		
		65	80	95
0	9·0	9·0	9·0	9·0
2	9·0	9·0	9·0	9·0
4	7·8**	9·0	9·0	9·0
6	6·2**	8·8	8·7	9·0
8	4·6**	8·8	8·8	8·8
12		8·0**	8·8	8·6
16		5·5**	8·5	8·8
21		4·0**	5·4**	8·5
27			4·7**	5·1**
32				4·5**

[a] Reproduced with permission from J. Fd Sci., 1977, 42(6), 1449. Courtesy of the Institute of Food Technologists.
[b] Flavour quality scale: 1 = extremely off-flavour, 2 = severely off-flavour, 3 = moderately off-flavour, 4 = slightly off-flavour, 5 = no fresh soybean curd flavours and no off-flavour (minimum acceptable quality), 6 = slightly fresh soybean flavour, 7 = moderate loss of fresh soybean curd flavour, 8 = slight loss of fresh soybean curd flavour, 9 = fresh soybean curd flavour.
** Significantly different from zero time sample at 1% level.

and core hardening prevented. Microwave processing[48,49] has been shown to be an ideal way of overcoming this potential problem and a commercial process has proved very successful. The pasta is first dried in a shaker drier to remove some water and so prevent clumping. This is followed by a 10-min exposure through a microwave chamber (915 MHz). Hot air is forced through the chamber to carry off the moisture liberated. Cooling, holding and packaging completes the process. The advantages are increased production and energy savings,[50,51] also a reduced floor area is required for the equipment. The time of rehydration and cooking of the product is reduced possibly because as the moisture escapes from the product during microwave drying, tiny tunnels are formed throughout the pasta. Total plate counts of bacteria are said to be much lower in the microwave product when compared to conventionally dried pasta.

4.5.2. Onions

A similar process to pasta drying has been applied to onions[50] and is being used commercially in California.

4.5.3. Field Corn

Experiments have been carried out using microwave energy (2450 MHz and 915 MHz) and untreated air to dry corn.[52] Conventional hot air drying causes core hardening of the kernels and the other disadvantage is the low thermal conductivity of the grain. Although the microwave process is feasible, there is a limit to the speed at which the corn may be dried and beyond this limit the swelling of the kernels would lower the market grade. It is thought that this may preclude the application of microwave power for drying corn.

4.5.4. Tomato Paste

Tomato paste foam has been dried by both hot air and microwave energy.[53,54] A disadvantage of drying a foamed product by hot air is the poor thermal conductivities of foams. Both methods yield a product of high quality but the microwave method (using microwave energy at 2450 MHz) reduces the drying time considerably.

4.5.5. Fungal Materials

A new protein food (fibrous fungal material) for human consumption is being developed jointly by Rank-Hovis McDougal (RHM) Ltd and the du Pont Company. *Fusarium graminearum* Schwabe is used and, after fermentation using a carbohydrate source, the resultant product requires a process to improve the texture and remove moisture. A combination of radio-frequency prepuffing and hot-air finishing has proved to be a reliable method. A wide range of frequencies from 27 MHz to 2450 MHz have been satisfactory.[55,56]

4.5.6. Japanese Foods

A puffing and drying process is used in Japan to produce puffed dried egg which is packed with noodles for fast service. Microwave drying is also used for tangle, scallop and noodle.[57]

4.5.7. Bacon

In North America precooked bacon is being produced commercially. Two processes have been described,[50,58] one company uses microwave heating and hot air while another one uses steam, hot air and microwaves. The water content of the bacon is reduced and it is claimed that the low temperature, short cooking time offers the potential for greatly reducing the formation of nitrosamines.[58,59] The product is sold for institutional use.

4.6. Freeze Drying

The application of microwaves in the freeze drying process has been discussed in a number of papers,[60-62] the main advantage being a reduction in the overall time of the process. A number of problems have been highlighted, one of them being electrical breakdown (corona discharge)[60,61] which has a deleterious effect on the product being dried.

Decareau[60] stressed the merits of a combined conventional/microwave freeze drying process where conventional freeze drying is used to remove the bulk of the moisture followed by microwave energy to remove the residual moisture. The problem is the design of the combination process.

A process has however been patented[63] which involves partially freeze-vacuum drying a food in particulate form to an average moisture content of 10–35 %. The frozen, partially freeze-vacuum dehydrated food is then subjected to microwave energy and the remaining core moisture is redistributed. The particles are then subjected to pressure to form a compressed food. The food processed could be uncooked vegetables or cooked meat.

Reports state that many of the problems of microwave freeze drying have been overcome,[61,62] but the process has not been adopted commercially.

4.7. Microwave Vacuum Drying

Recently two applications of microwave vacuum drying have been reported; one machine, GIGAVAC (Industries Micro-Ondes Internationales (IMI), Epone, France), is currently being used to produce an orange powder commercially.[64,65] The other, MIVAC (McDonnell Douglas Corporation, USA), is a prototype which has been developed for drying grain.[66] The process for drying the orange concentrate is similar to vacuum drying processes used for tea and other beverages, the difference being that microwave energy is used instead of radiant heaters. It is claimed that the cost per kilogramme of product is much lower than freeze drying and even lower than spray drying.

Quality advantages described for the orange powder produced from a preconcentrate juice are retention of the typical orange flavour, aroma and colour, which is possibly due to the temperature of processing being low (maximum 40 °C) and the dehydration times being short.[65] It is claimed that because the grain temperature is kept low stress cracking is prevented.[66]

The following range of products have been successfully tested on an experimental basis:[66] Corn, peanuts, rice, wheat, bananas, raisin grapes, peaches, apricots, strawberries, raspberries, apples, peppers, tomatoes, soy protein and milk powder.

4.8. Thawing

4.8.1. Butter

Large blocks (approximately $0.3 \, m^3$ packed in cardboard boxes) of butter are thawed by passing them through a microwave heating chamber for 15 min.[32] A 90 kW, 915 MHz machine processes 1100 kg of butter per hour and no surface heat facility is required. On average the temperature rise is from -20 to $+2$ °C.

4.8.2. Meat

Partial thawing or tempering has been described by Bezanson[67] and Nott.[68] Dicing, cutting and mincing meat is usually carried out at a temperature of between -3 to -2 °C. A microwave tempering tunnel enables blocks of meat to be taken straight from the cold store so that it enters the microwave unit at -20 °C, and after 0.5 2 min leaves the microwave unit at a temperature of approximately -3 °C. The tempering process, conventionally, is carried out in conditioning rooms and takes 48–72 h. In the UK, microwave equipment at a frequency of 896 MHz $\pm$ 10 MHz is used and there are various types of tunnels (ABR Food Machinery International Ltd, Bletchley, Milton Keynes, UK) ranging from a throughput of 1000 kg/h to 3350 kg/h which can be used for tempering meat, fish and poultry for medium- to large-scale processes.

During thawing, as the temperature of beef reaches 1 °C, major changes occur in the dielectric properties of the beef resulting in a marked decrease in the penetration of microwave energy which often results in 'runaway heating'.[69] For this reason experiments have been carried out on the microwave thawing of tuna fish, beef and pork using a surface cooling process—electrostatic spraying of cryogenic liquids.[70]

An evaluation done at the US Army Natick Research Laboratories showed the use of microwave tempering of meat to be a useful operation in central food preparation systems.[71]

5. MICROWAVE COOKERS

The basic principles of microwave cookers have been described fully by Van Zante,[72] Osepchuk[73] and Barber.[74] It is usual for consumer microwave cooking appliances to operate at a frequency of 2450 MHz.

The domestic microwave cooker market in the UK is predicted to reach a market saturation level of 10% by 1984 and a constant market level of at least half a million cookers per annum by 1985.[75]

Three configurations of microwave cookers exist,[76] counter top models which are the most popular, double ovens (i.e. a conventional and a microwave cooker mounted together)[77] and combination microwave/conventional ovens. In the UK a combined forced air convection and microwave oven is marketed by Mealstream (UK) Ltd. The Mealstream 605 is too large and expensive for the domestic market but it is very useful and successful in the food service industry.

5.1. Methods of Distributing Microwave Energy

The roof of the interior of the cooking space of the microwave cooker usually incorporates a mode stirrer, this may be covered by a spatter shield. The function of the stirrer is to attempt to distribute evenly within the cooking space the microwaves from the magnetron.[78] Some cookers have a turntable which replaces the stirrer although the Belling MW1 has both. Experiments to improve the heating uniformity in microwave ovens are continually being carried out. Kashyap and Wyslonzil[79] described a multi-slot waveguide oven feed suitable for domestic microwave ovens. Other experiments they described involve sweeping the frequency over a certain range.

The turntables are usually driven by a motor which is coupled by gears to a rotary shaft which extends into the cooking space through the bottom plate. A patent exists for a rotary table with a magnetic driving system,[80] thus eliminating the necessity of the rotary shaft extending into the cooking space. A portable battery driven turntable suitable for use in microwave cookers has also been developed.[81]

In spite of these attempts to distribute the energy evenly in the cavity in practice the energy distribution, even when the cooker is empty, is not uniform. Many tests designed to measure the electric field distribution in a microwave cooker have been described.[82,83] The aim of these experiments being, if possible, to enable the user to utilise the hot and cold spots appropriately. The lack of uniformity in the distribution of energy in the cooking space is a major problem in the utilisation of microwaves for heating food.

5.2. Cooking Power Output

The power available for cooking in the majority of microwave cookers ranges from 72 to 720 W (see Table 6) which shows a comparison of a number of counter top microwave cookers available in the UK.

It can be seen from Table 6 that the output power into the cavity can be varied in some cookers, e.g. Belling MW1 has two output settings whereas the Toshiba ER798 has a variable output from 72 to 720 W in 99 stages.

TABLE 6

A SELECTION OF MICROWAVE COOKERS ON THE UK MARKET. ALL OPERATE AT 240 V AND A FREQUENCY OF 2450 MHz

Model	Input (W)	Output (W)	Other features
Belling MW1	1 300	260 or 700	Motor driven automatic turntable and two power settings
Moffat 4001	1 450	650	Settings 1–6 cycle power on and off at varying rates
Toshiba ER798	1 370	Variable 72–720 (99 stages)	Heat sensor probe. Auto-defrost (automatic change of power levels during defrost to ensure complete defrosting). Programmable cooking, cook by time then heat/hold
Sharp R-8200E	1 400	Microwave 600, browning 1 300	Turntable (no mode stirrer). Settings 1–4 cycle power on and off at varying rates. Infrared browner can be programmed to switch on automatically at the end of the microwave cooking time and will brown food
Creda	1 500	Variable 100–700	Settings 1–6 cycle power on and off at varying rates (12 s pulsing action). Browning tray supplied

Many cookers can also cycle the power on and off. The ratio of the on : off time can be controlled by a simple switching arrangement. This facility is extremely useful for defrosting frozen food[84] and also for slowing down the cooking process.[78]

5.3. Temperature Sensing Devices and Microprocessors

Some of the more expensive microwave cookers are equipped with a device to sense the internal temperature of the cooked food. To date the most common method has been the inclusion of a temperature probe.[78,85] It is usual for the temperature sensor to be set to cook the food to a selected internal temperature and then shut the oven off or the cooker may have a 'keep-warm' or 'heat/hold' setting which can be selected to operate at this time.

Buck[86,87] has described a method of cooking food, e.g. meat[88] by sensing *in situ* humidity and temperature conditions of the microwave cooker cavity. The humidity sensor and the temperature sensor are positioned adjacent to and by the exterior of the ventilation part of the microwave cooker cavity. The sensors are connected to a programmable controller and input the sensed *in situ* environmental conditions to the programmable controller. The memory of the programmable controller stores 'characteristic humidity curves' for microwave cooking of different types of foods. A microwave cooking algorithm can be selected, a numeric keyboard being connected to the programmable controller.

Although electronic controls are more expensive than electromechanical methods Winstead[89] maintains that they offer a significant improvement in the performance of the microwave cooker.

5.4. Lack of Standardisation
The existing cooking power settings on microwave cookers from different manufacturers do not equate. This major problem has been well explained by Katz[90] and Schiffman.[91] It makes it impossible for a general guide to be written for food preparation using microwaves. It is also a major problem in the food industry regarding the labelling of manufactured products with microwave cooking instructions.

Attempts are being made to standardise the method of measuring the power output of microwave cookers, followed by the adoption of a standard method of labelling[92] (see the settings as shown on Table 7). A committee has been established by the International Microwave Power

TABLE 7
IMPI-CAS RECOMMENDED LABELLING OF THE CONTROL SETTINGS ON MICROWAVE COOKERS[92]

Power level setting designation	Power output at setting (W)	Percentage of high-setting
High	550 ± 50	100
Medium high	385 ± 40	70
Medium	275 ± 30	50
Medium low	165 ± 20	30
Low	55 ± 15	10

Institute–Cooking Appliance Section (IMPI–CAS) which plans to send these recommendations to all concerned parties including the British Standards Institution.

6. FOOD PRODUCT DEVELOPMENT AND MICROWAVES

There are some products, e.g. precooked pancakes with margarine syrup and fruit, popcorn, etc.,[93] which have been developed in the USA, specifically for heating in a microwave oven. A patent exists for the special arrangement in a package of bread, a filling and condiment for microwave heating.[94]

Research continues for packaging materials which can be used both in microwave cookers and conventional ovens.[95,96]

Again in North America it is reported that food companies are developing flavours and condiments to replace missing notes, e.g. browning butter, roasting beef, etc. Baking mixes for microwave cooking are beginning to appear, e.g. a crumb topping in which it is recommended that brown butter and caramelised flavours are included in the product.[97]

Decareau[98] lists 20 product development ideas for the microwave cooker owner. Two good reasons for the growth of this market are given

1.   The increasing sale of microwave cookers
2.   Energy savings; usually less energy is required for microwave cooking, figures of between 62 to 75 % have been quoted.[84,98,99]

7. THE EFFECT OF MICROWAVE COOKING ON FLAVOUR

7.1. Meat
MacLeod and Coppock[100] investigated the flavour of beef cooked conventionally and by microwave energy. The odour volatiles were analysed by using gas chromatography/mass spectrometry. The separated odour volatiles were then evaluated by sensory analysis. For both conventional and microwave heating methods samples were analysed in the presence and absence of water. Their results indicated that for both the boiled and roast series of samples, cooked to an acceptable and equivalent degree of doneness by conventional and microwave cooking produced different amounts of volatiles. The microwave cooked beef samples liberated approximately one half of the total volatiles of the conventional

product. Thus they stated that it would appear that the production of flavour volatiles is not purely thermal and is to some extent dependent upon time. There was an initial increased rate of production of volatile flavour components, due to the more rapid rise in temperature in the microwave heated samples but this was not sufficient to make up for the shorter cooking time when the total production of volatiles was considered.

In further work on meat[101] in which 100 volatile components were isolated it was found that in general the classes of components which represent the greater proportion of isolates for the conventional boiled beef were the benzenoids, aldehydes and furans, whereas the classes of compounds which in general occur in greater amounts in the microwave boiled beef isolates were the alkanes, alkenes, alcohols and pyrazines.

It is of interest to note that several of the alkanes, alkenes and alcohols present in high concentrations in the microwave boiled beef aroma were associated with relatively undesirable odour assessments such as sour, harsh, pungent, unpleasant, dull cardboard. The conventionally boiled beef aroma was described in more desirable terms such as pleasant, solvent, fruity and meaty. More evidence was presented by MacLeod and Coppock[102] substantiating that relatively undesirable odour qualities are present in microwave samples cooked for the normal time required for cooking the meat by either boiling or roasting. The same authors point out that chemical reactions in general are strongly temperature dependent and the rapid initial temperature rise in microwave cooking has to be contrasted with the slower conventional process of conduction where a high temperature gradient initially exists throughout the food. The compounds present in higher concentrations in the microwave samples are probably representative of compounds formed by reactions which have relatively high rates of reaction and are thus formed in significant amounts in the relatively short microwave cooking time. They concluded from this work that time appears to be an important factor as far as desirable beef flavour production is concerned. The microwave samples were cooked by a continuous microwave method (2450 MHz).

Ream and co-workers[103] also compared conventional and continuous microwave cooking. Small (1362 g) and large (3·63–4·08 kg) beef roasts were used; highly significant differences in all areas of sensory evaluation were found. The meat cooked by the microwave method was less tender, juicy and flavourful according to the sensory panel and off-flavour was commented on more often in the microwave cooked meat samples.

These findings therefore substantiate the wisdom of the development of microwave cookers with timing mechanisms to slow down the rate of heating and recently a number of papers have been published comparing intermittent microwave cooking methods with conventional cooking.[104-106] With one exception[104] all reported that there were no significant differences in the scores for flavour of meat cooked by both methods when evaluated by a sensory panel. Korschgen *et al.*[104] showed a significant difference in favour of conventional cooking when the edge of pork and lamb slices were compared (see Table 8). This seemed to be due to the greater degree of browning which occurs with conventional cooking.

TABLE 8

MEAN SENSORY SCORES FOR ROASTS COOKED BY MICROWAVE AND CONVENTIONAL METHODS[a]

Species and treatment	Flavour score[b] for sample from interior slice	Flavour score[b] for sample from edge of slice
Beef		
Microwave 220 V	3·6	3·5
Microwave 115 V	3·2	3·2
Conventional	3·4	3·4
Pork		
Microwave 220 V	3·0	2·9
Microwave 115 V	2·8	2·7 ⎫ *
Conventional	3·2	3·2 ⎭
Lamb		
Microwave 220 V	3·3	3·9
Microwave 115 V	3·2	3·5 ⎫ *
Conventional	3·6	4·2 ⎭

[a] Reproduced with permission from *J. Am. diet. Ass.*, 1976, **69**, 635. Courtesy of The American Dietetic Association.
[b] Number of scores = 35. Range of scores for flavour: 5 = desirable, 1 = undesirable.
* Significantly different from each other ($p < 0.05$).

Starrack[107] commented on the improvement of roast meat when an intermittent method of microwave cooking is used, details of methods for roasting are also given.

Meat roasting has been a subject of research at Natick Development Center for many years. Decareau[60] has recommended that because of the temperature pattern in beef roast cooked by microwaves (see Fig. 2) the roast should be removed from the cooker before the internal recommended temperature was reached. Time should then be allowed for conduction to

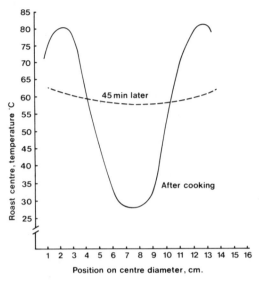

FIG. 2.   Temperature pattern in beef roast cooked by means of microwave energy at 2450 MHz (from Decareau[60]).

occur and the temperature equilibrate. The timing devices on some cookers which control intermittent microwave cooking do allow some conduction to take place during the cooking cycle.

Recent work at Natick[108] aims to optimise the process of the roasting meat in terms of yield, quality and cooking time. Uniform, cylindrical pieces of meat are being used (12 cm by 20 cm). First of all conventional roasting studies were carried out carefully monitoring changes in weight with time and also collecting and measuring drip losses so that a figure for evaporative cooking could be worked out. Dielectric properties of beef as a function of temperature and microwave frequency were obtained and a computer program was developed which can provide a graphical print-out of the temperature in a cylindrical roast of beef given specific cooking conditions.

To verify the computer studies an experimental oven was built which has variable microwave power at both 915 MHz and 2450 MHz to approximately 2 kW, controlled oven temperature from 93 to 260 °C, controlled steam pressure at 5, 10 and 15 psig and an IBM card reader to permit programmes to be run in an automatic mode.

The results of the computer program have demonstrated good agreement with limited experimental data.[108] Other results have shown that

2450 MHz microwave power was inferior to 915 MHz for roasting beef cylinders with diameters greater than 6 cm. Indications so far are that the following methods are superior: 300 W at 915 MHz, or a combination of 600 W at 915 MHz for the first 20 min followed by 300 W at 915 MHz.

7.2. Vegetables

A detailed investigation of the flavour volatiles of cooked cabbage has been carried out.[109] First of all tasting tests were done to find out a cooking time which provided the product with the best flavour conventionally. This was found to be cabbage cooked for ten minutes and was referred to as 'ideally' cooked cabbage.

The flavour volatiles liberated were collected and analysed by gas chromatography and mass spectrometry. The relative percentage abundance of each volatile was calculated and the values compared for the different cooking variations. Approximately 30 volatiles were isolated after 10 min conventional cooking and these were used as a baseline to be compared with the volatiles isolated by other cooking procedures. One of these was cooking by microwave energy (2450 MHz) for varying periods of time in the presence and absence of water.

The results obtained for both wet and dry microwave cooking resembled each other but were both different from the conventional method, i.e. the ideal. One of the differences was a marked increase in methyl alcohol during microwave heating. This increase occurred sooner during dry microwave heating as opposed to wet microwave heating. A similar effect was also observed for allyl isothiocyanate, thus showing the more rapid action of microwaves when no added water is present.

Non-volatile organic acids (lactic, succinic, malic and citric) have been identified and quantified in frozen peas cooked by microwaves.[110] Gas–liquid chromatography was used for this study. The concentration of these acids increased after cooking and in particular samples cooked by microwaves without water at 115 V and 550 W cooking power. Malic acid was identified in frozen carrots cooked in the same way but in this vegetable concentration was not affected by cooking method. The non-volatile organic acids found in both peas and carrots were not correlated with sensory scores for flavour.

Mabesa and Baldwin[111] carried out similar cooking procedures as mentioned in ref. 110. Their findings showed that the sensory scores for the flavour of peas cooked with and without water in a consumer microwave cooker (115 V, 550 W) were not significantly different from each other and that all the scores were in the acceptable range. A difference was found, however, between peas cooked in microwaves at 115 V (550 W cooking power) and those cooked at 220 V (1150 W cooking power). Significantly higher scores for flavour were obtained by the 115 V, 550 W method. Higher flavour scores were obtained for carrots cooked by microwaves without water when compared to the other methods.

The flavour scores for microwave cooked peas and carrots were either not significantly different from the conventionally cooked counterpart or were rated significantly higher. The authors[111] stress, however, that caution must be exercised in generalising as to the effects of microwave cooking on vegetables. They noted differences in trends for peas and carrots. They also commented on biological variation within and between lots of vegetables.

These results, however, do indicate that cookers in which the power can be varied could be beneficial for the flavour of certain products.

7.3. Baked Products
The results of a comparison of sensory scores for flavour of bread baked in a microwave cooker (2450 MHz, 220 V, 2000 W) compared to a convection oven are shown on Table 9.[112] The flavour was considered to be better on the second day after baking and this was true for both methods of cooking. The poorer flavour scores for the microwave baked whole-wheat bread were not unexpected and were attributed to the lack of caramelisation and

TABLE 9

FLAVOUR SCORES FOR BREAD BAKED IN MICROWAVE AND CONVECTION OVENS[a]

Type of bread	Oven	No. of days storage	Flavour score[b]
Whole wheat	Microwave	1	2·3
		2	2·2
	Convection	1	1·8
		2	1·6
Rye	Microwave	1	2·9
		2	2·9
	Convection	1	3·0
		2	2·9

[a] Reproduced with permission from Fd Technol., 1973, 27(12), 34. Courtesy of the Institute of Food Technologists.
[b] Average of two evaluations done by 20 assessors. Flavour scale: 1 = like, 2 = like slightly, 3 = undecided, 4 = dislike slightly, 5 = dislike.

development of flavour compounds which occur during conventional baking. Some of the panel members commented that the microwave bread had a very mild flavour or less flavour than the bread baked in the convection oven.

With the rye bread however the panels did not detect any difference in flavour between the two baking methods.

8. CONCLUSION

The continued expansion and improvement of the application of microwaves to food requires integration of the basic knowledge of the composition and heating parameters of food with the technology of food processing. The complexity of many foods and biological variation do not assist the application of microwaves. The collection of further data on conventional, as well as microwave processing and the use of the computer may help to solve some of these problems.

Even distribution of microwave energy and the standardisation of settings on microwave cookers would greatly assist both the food manufacturers and the users of microwave cookers. It seems a contradiction that the microwave cooker, an appliance of a scientific age, often requires the development of an art to use it! Standardisation of the cookers followed by the standardisation of cooking procedures in a scientific manner would enhance consumer acceptance for those without inclination to develop the art. Improvements are being continually made to achieve these aims and the sophistication of the recent cookers enhance their versatility.

In 1976 Lorenz,[113] in a comprehensive review paper, commented on the lack of published data which directly compared the qualitative and quantitative chemical composition of flavour components of microwave and conventionally heated foods. The importance of work in this area was stressed, flavour being one of the most important properties of the resultant product. It is encouraging to see that some progress since 1976 has been made in this area of research and further expansion should be seen in the future.

REFERENCES

1. BENGTSSON, N. E. and RISMAN, P. O., *J. Microwave Power*, 1971, 6(2), 107.
2. RISMAN, P. O. and BENGTSSON, N. E., *J. Microwave Power*, 1971, 6(2), 101.

3. ROEBUCK, B. D., GOLDBLITH, S. A. and WESTPHAL, W. B., *J. Fd Sci.*, 1972, **37**(2), 199.
4. TINGA, W. R. and NELSON, S. O., *J. Microwave Power*, 1974, **8**(1), 23.
5. TO, E. C., MUDGETT, R. E., WANG, D. I. C., GOLDBLITH, S. A. and DECAREAU, R. V., *J. Microwave Power*, 1974, **9**(4), 303.
6. MUDGETT, R. E., SMITH, A. C., WANG, D. I. C. and GOLDBLITH, S. A., *J. Fd Sci.*, 1971, **36**, 915.
7. OHLSSON, TH. and BENGTSSON, N. E., *J. Microwave Power*, 1975, **10**(1), 93.
8. METAXAS, A. C., In *Industrial Applications of Microwave Energy* (Smith, R. B., Ed.), Transactions of the International Microwave Power Institute Vol. 2, IMPI, New York and London, 1976, p. 19.
9. OHLSSON, TH. and RISMAN, P. O., *J. Microwave Power*, 1978, **13**(4), 303.
10. MUDGETT, R. E., GOLDBLITH, S. A., WANG, D. I. C. and WESTPHAL, W. B., *J. Microwave Power*, 1980, **15**(1), 27.
11. GERLING, J. E., *J. Microwave Power*, 1978, **13**(1), 37.
12. COPSON, D. A., *Microwave Heating*, 2nd edn, AVI Westport, Connecticut, 1975 p. 2.
13. KAMAT, G. P. and LASKEY, J. W., Radiation Bio-effects, Summary Report, US Dept. of Health, Education and Welfare, Publ. No. BRH/DBE, 70, 1970.
14. LANGLEY, J. B., YEARGERS, E. K., SHEPPARD, A. P. and HUDDLESTON, G. K., 'Proceedings 1973 Microwave Power Symposium' IMPI Loughborough, England, 1973.
15. BINI, M., CHECCUCCI, A., IGNESTI, A., MILLANTA, L., RUBINO, N., CAMICI, G., MANAO, G. and RAMPONI, G., *J. Microwave Power*, 1978, **13**(1), 95.
16. COPSON, D. A., US Patent 2,833,657, 1958.
17. KENT-JONES, D. W. and AMOS, A. J., *Modern Cereal Chemistry*, 6th edn, Food Trade Press Ltd, London, 1967, p. 368.
18. AREF, M. M., NOEL, J-G. and MILLER, H., *J. Microwave Power*, 1972, **7**(3), 215.
19. AVISSE, C. and VAROQUAUX, P., *J. Microwave Power*, 1977, **12**(1), 73.
20. MATHEW, A. G. and PARPIA, H. A. B., *Advances in Food Research 19*, (Chichester, C. O., Mrak, E. M. and Stewart, G. F., Eds.), Academic Press, Inc. London, 1971, p. 75.
21. MAPSON, L. W., SWAIN, T. and TOMALIN, A. W., *J. Sci. Fd Agr.*, 1963, **14**, 673.
22. COLLINS, J. L. and McCARTY, I. E., *Fd Technol.*, 1969, **23**, 337.
23. CHEN, S. C., COLLINS, J. L., McCARTY, I. E. and JOHNSTON, M. R., *J. Fd Sci.*, 1971, **36**, 742.
24. HILL, M. A. and PILL, C. M. *Proceedings XIV Symposium International sur les Applications Energétique des Micro-ondes*, Monaco. Comité Français d'Electrothermie, Paris, 1979, p. 55.
25. ABARA, A. E. and HILL, M. A., *J. Microwave Power*, in press.
26. HUXSOLL, C. C., DIETRICH, W. C. and MORGAN, JR, A. I., *Fd Technol.*, 1970, **24**(3), 84.
27. DIETRICH, W. C., HUXSOLL, C. C., WAGNER, J. R. and GUADAGNI, P. G., *Fd Technol.*, 1970, **24**, 293.
28. EHEART, M. S., *J. Am. diet. Ass.*, 1967, **50**, 207.
29. NOBLE, I. and GORDON, J., *J. Am. diet. Ass.*, 1964, **44**, 120.
30. SWEENEY, J. P., GILPIN, G. L., MARTIN, M. E. and DAWSON, E. H., *J. Am. diet. Ass.*, 1960, **36**, 122.

31. DIETRICH, W. C., HUXSOLL, C. C. and GUADAGNI, D. G., *Fd Technol.*, 1970, **24**, 613.
32. MEREDITH, R. J. In *Techniques and Applications of Microwave Power*, Digest No. 1979/65, Institution of Electrical Engineers, Electronics Division Professional Group E11 (Antennas and Propagation), London, 1979, p. 8.1.
33. MEREDITH, R. J., Personal communication, 1980.
34. SMITH, F. J. and WILLIAMS, L. G., US Patent 3,578,463, 1971.
35. OHLSSON, Th., *How Ready are Ready-to-Serve Foods?* (Paulus, K., Ed.), S. Karger, Basel, 1977, p. 105.
36. KENYON, E. M., WESTCOTT, D. E., LA CASSE, P. and GOULD, J. W., *J. Fd Sci.*, 1971, **36**, 289.
37. AYOUB, J. A., BERKOWITZ, D., KENYON, E. M. and WADSWORTH, C. K., *J. Fd Sci.*, 1974, **39**, 309.
38. STENSTRÖM, L. A., 'IMPI Symposium', Ottawa, Canada, 24–26 May 1972, Paper 7.4.
39. O'MEARA, J. P., TINGA, W. R., WADSWORTH, C. K. and FARKAS, D. F., *J. Microwave Power*, 1976, **11**(2), 213.
40. HECHT, J. L. and HASKELL, V. C., *J. Microwave Power*, 1976, **11**(2), 211.
41. LIN, C. C. and LI, C. F., *J. Microwave Power*, 1971, **6**(1), 45.
42. SALE, A. J. H., *J. Fd Technol.*, 1976, **11**(4), 319.
43. Unilever, G.B. Patent 1,187,766, 1970.
44. BENGTSSON, W. E., GREEN, W. and DEL VALLE, F. R., *J. Fd Sci.*, 1970, **35**(5), 681.
45. HAMID, M. A. K., BOULANGER, R. J., TONG, S. C., GALLOP, R. A. and PEREIRA, R. R., *J. Microwave Power*, 1969, **4**(4), 272.
46. WU, M. T. and SALUNKHE, D. K., *J. Fd Sci.*, 1977, **42**(6), 1448.
47. MCMILLAN, D. C., US Patent 3,615,726, 1971.
48. MAURER, R. L., TREMBLAY, M. R. and CHADWICK, E. A., *Fd Technol.*, 1971, **25**(12), 1244.
49. Unilever N.V., The Netherlands Patent Application 7,106,629, 1971.
50. SMITH, F. J., *Microwave Energy Applications Newsletter*, 1979, **XII**(6), 6.
51. ANON., *Fd Engng*, 1972, **44**(4), 94.
52. FANSLOW, G. E. and SAUL, R. A., *J. Microwave Power*, 1971, **6**(3), 229.
53. RZEPECKA-STUCHLEY, M. A., BRYGIDYR, A. M. and McCONNELL, M. B., *J. Microwave Power*, 1976, **11**(2), 215.
54. BRYGIDYR, A. M., RZEPECKA, M. A. and McCONNELL, M. B., *Canadian Institute of Food Science and Technology*, 1977, **10**(4), 313.
55. HUANG, H. F. and YATES, R. A., *J. Microwave Power*, 1980, **15**(1), 15.
56. HUANG, H. F. *Proceedings XIV Symposium International sur les Applications Energétiques des Micro-ondes*, Monaco. Comité Français d'Electrothermie, Paris, 1979, p. 66.
57. KASE, Y. and OGURA, K., *J. Microwave Power*, 1978, **13**(2), 115.
58. ANON., *Food Engineering International*, 1977, **2**(8), 38.
59. MATTSON, P., *Food Product Development*, 1978, **12**(4), 47.
60. DECAREAU, R. V. (Ed.), In *Microwave Ovens and Frozen Foods Make Cents*, Transactions of the International Microwave Power Institute Vol. 5, IMPI, New York and London, 1975, p. 101.
61. PETTRE, R. P., ARSEN, H. B. and MA, Y. H., *AICHE Symposium Series*, 1977, **73**(163), 131.

62. ANG, T. K., FORD, J. D. and PEI, D. C. T., *J. Fd Sci.*, 1978, **43**(2), 648.
63. RAHMAN, A. R., US Patent 4,096,283, 1978.
64. ANON., *Microwave Energy Applications Newsletter*, 1979, **XII**(3), 12.
65. ATTIYATE, Y., *Food Engineering International*, 1979, **4**(1), 30.
66. ELIAS, S., *Food Engineering International*, 1979, **4**(1), 32.
67. BEZANSON, A. F., In *Microwave Ovens and Frozen Foods Make Cents* (Decareau, R. V., Ed.), Transactions of the International Microwave Power Institute Vol. 5, IMPI, New York and London, 1975, p. 89.
68. NOTT, B., *Fd Process Ind.*, 1977, **46**(548), 18.
69. DECAREAU, R. V., *Microwave Energy Applications Newsletter*, 1975, **8**(2), 3.
70. BIALOD, D., JOLION, M., LEGOFF, R., *J. Microwave Power*, 1978, **13**(3), 269.
71. SWIFT, J. and TUOMY, J. M., *Microwave Energy Applications Newsletter*, 1978, **XI**(1), 3.
72. VAN ZANTE, H. J. *The Microwave Oven*, Houghton Mifflin Co. Boston, 1973.
73. OSEPCHUK, J. M. In *Future Impact of Microwave Ovens on the Food Industry* (McConnell, D. R., Ed.), Transactions of the International Microwave Power Institute Vol. 4, IMPI, New York and London, 1975, p. 5.
74. BARBER, H., In *Microwave a Cooking Revolution*, Graham Kemp Associates, London, 1979.
75. GILES, P. G., In *Microwave a Cooking Revolution*, Graham Kemp Associates, London, 1979.
76. RENNEKAMP, R., *Microwave Energy Applications Newsletter*, 1978, **XI**(6), 7.
77. NIKOLOV, M., *Microwave World*, 1980, **1**(1), 12.
78. CONACHER, G. and WEBB, J., *Microwave Cooking at Home*, The Electricity Council, London, EC 3991/5.80, 1980.
79. KASHYAP, S. C. and WYSLONZIL, W., *J. Microwave Power*, 1977, **12**(3), 223.
80. TANAKA, J. and KAI, T., US Patent 4,163,141, 1979.
81. BOWEN, B., *Microwave World*, 1980, **1**(1), 14.
82. RINGLE, E. C. and DONALDSON DAVID, B., *Fd Technol.*, 1975, **29**(12), 46.
83. BOBENG, B. and DONALDSON DAVID, B., *Microwave Energy Applications Newsletter*, 1975, **9**(6), 3.
84. SCHIFFMAN, R. F. In *Microwave a Cooking Revolution* Graham Kemp Associates, London, 1979.
85. BURBICK, D. C. and RENNEKAMP, R. G., *J. Microwave Power*, 1978, **13**(1), 3.
86. BUCK, R. G., US Patent 4,162,381, 1979.
87. BUCK, R. G., US Patent 4,154,855, 1979.
88. BUCK, R. G., US Patent 4,171,382, 1979.
89. WINSTEAD, D., *J. Microwave Power*, 1978, **13**(1), 7.
90. KATZ, M. H., *Food Product Development*, 1977, **11**(9), 48.
91. SCHIFFMAN, R. F., *Food Product Development*, 1979, **13**(7), 38.
92. ANON., *Microwave World*, 1980, **1**(2), 8.
93. MOORE, K., *Food Product Development*, 1979, **13**(7), 36.
94. STANDING, C. N. and BRANDBERG, L. C., US Patent 4,133,896, 1979.
95. BOUTIN, R. J., *J. Microwave Power*, 1978, **13**(1), 47.
96. COOK, R. E. In *Consumer Microwave Oven Systems Conference* Association of Home Appliance Manufacturers, Chicago, USA, 1972.
97. ANON., *Food Product Development*, 1978, **12**(10), 32.
98. DECAREAU, R. V., *Microwave Energy Applications Newsletter*, 1975, **8**(1), 3.

99. McCONNELL, D. R., *J. Microwave Power*, 1974, **9**(4), 341.
100. MacLEOD, G. COPPOCK, B. M., *Proc. Int Congress Fd Sci. and Technol.*, Spain, 1974, **1**, 6.
101. MacLEOD, G. and COPPOCK, B. M., *J. Agric. Fd Chem.*, 1976, **24**(4), 835.
102. MacLEOD, G. and COPPOCK, B. M., *J. Fd Sci.*, 1978, **43**, 145.
103. REAM, E. E., WILCOX, E. B., TAYLOR, F. G. and BENNETT, J. A., *J. Am. diet. Ass.*, 1974, **65**(2), 155.
104. KORSCHGEN, B. M., BALDWIN, R. E. and SNIDER, S., *J. Am. diet. Ass.*, 1976, **69**, 635.
105. HAWRYSH, Z. J., PRICE, M. A. and BERG, R. T., *Canadian Institute of Food Science and Technology*, 1979, **12**(2), 78.
106. KORSCHGEN, B. M. and BALDWIN, R. E., *J. Microwave Power*, 1978, **13**(3), 257.
107. STARRACK, G., *The Microwave Energy Applications Newsletter*, 1979, **XII**(4), 15.
108. NYKVIST, W. E. and DECARFAU, R., *J. Microwave Power*, 1976, **11**(1), 3.
109. MacLEOD, A. J. and MacLEOD, G., *J. Fd Sci.*, 1970, **35**, 744.
110. MABESA, L. B., BALDWIN, R. E. and GARNER, G. B., *J. Food Protection*, 1979, **42**(5), 385.
111. MABESA, L. B. and BALDWIN, R. E., *J. Microwave Power*, 1978, **13**(4), 321.
112. LORENZ, K., CHARMAN, E. and DILSAVER, W., *Fd Technol.*, 1973, **27**(12), 28.
113. LORENZ, K., *CRC Critical Reviews in Food Science and Nutrition*, 1976, **7**(4), 339.

Chapter 5

FREEZE DRYING: THE PROCESS, EQUIPMENT AND PRODUCTS

J. Lorentzen

A/S Atlas, Copenhagen, Denmark

SUMMARY

Freeze drying of foods is normally carried out by prefreezing in a separate freezer and by sublimation drying and secondary drying in a vacuum cabinet with heater plates. The main physical processes of the sublimation drying are heat transfer from the heater plates to the sublimation front and water vapour transport from the sublimation front to the ice condenser. A mathematical treatment gives an insight into the process, but a true picture can only be found by laboratory tests. Such tests are also necessary to establish process parameters for each product. The main features of normal freeze drying equipment are given, and some characteristics of freeze dried products are discussed.

1. THE FREEZE DRYING PROCESS

The common illustration of the freeze drying process is a time–temperature diagram giving the temperature history of the product.[1] In this context the freeze drying process is subdivided into different steps, and usually three main steps are considered: the prefreezing, the sublimation drying and the secondary drying.

In the prefreezing step the product is chilled to normal deep freeze temperature. The major part of the water content will freeze out as ice crystals, and the total water content is immobilised in its location within the product.

In sublimation drying the ice crystals are sublimated into water vapour at the low sublimation temperature and water vapour removed from the product.

In secondary drying 'bound water', which did not freeze out as ice crystals, is reduced to the low level desired in the product by raising the temperature conditions.

This time–temperature representation describes the freeze drying process in each microscopic element of the product. The various elements of each product piece will run through different process cycles of this kind. In macroscopic product pieces the prefreezing will normally run faster at the surface than in the interior of the product, and the sublimation drying of interior zones will overlap the secondary drying of the superficial layers.

Let us regard one piece of food about halfway through the freeze drying process in a vacuum freeze drier (Fig. 1).[2] Heat radiation hits the top

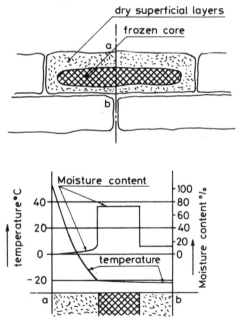

FIG. 1.   A piece of food during freeze drying. Heat input by radiation to top surface. Momentary variation in moisture content and in temperature along section a–b.

surface of the product piece, which is surrounded by similar product pieces on all other surfaces. We can distinguish between the still frozen core of this piece and the superficial layers, where ice crystals no longer exist. The boundary of the frozen core is known as the sublimation front and is playing a prominent part in the freeze drying process.

The microscopic elements forming the sublimation front are running through the sublimation process step. The product elements inside the core are still in the prefreezing step, as they have not yet started the sublimation drying. The product elements of the superficial layers are now in the secondary drying period. Figure 1 illustrates the momentary process situation for each element along a section a–b through the product piece. The situation is characterised by the local product temperature and by the local moisture content. Note that the local advance of the secondary drying step is linked with the local temperature rise.

At the sublimation front the major part of the water content sublimes from ice crystals to water vapour. This consumes the major part of the heat input of the process. This heat energy has to be carried to the sublimation front, and the water vapour produced has to be transported away from the sublimation front. As a quantity of product ice disappears, the sublimation front moves slightly towards the product core.

These two main transport mechanisms within the product, the heat transport to the sublimation front and the vapour transport from the sublimation front, are directly coupled. In each element of the sublimation front the temperature corresponds to the water vapour pressure, the saturation pressure over ice. The driving force of the water vapour movement is the pressure differential between the water vapour at the sublimation front and the water vapour in the freeze drying cabinet at the product surface. The driving force for this heat transfer is the difference between the temperature at the surface of the product piece and the temperature at the sublimation front.

By temperature measurements at defined locations in the product during the freeze drying process, as shown in Fig. 2, a good picture of the advancing sublimation front and of the local progress of the secondary drying can be obtained.[3]

Another representation of the process is given by the temperatures in defined locations as a function of residual moisture.[4]

When disregarding the secondary drying, the sublimation drying process within the product can be regarded as a combination of only two basic transfer processes, the heat conduction between the product surface and the sublimation front and the vapour flow from the sublimation front to the

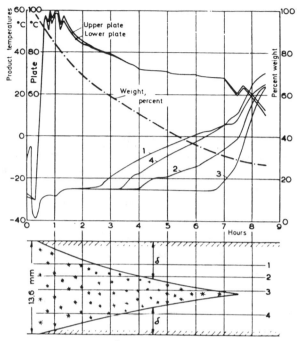

FIG. 2. Temperatures and weights of slices of raw meat during the freeze drying
process. Lower diagram indicates sublimation front movement.

product surface. Process wise this is much simpler than an ordinary drying
process, where the movement of a liquid phase has also to be considered.

Based on this and other simplifying assumptions mathematical models
for the freeze drying process have been given for different product
configurations.[5] Such solutions give valuable insight into the nature of the
process, and in combination with experimental work they have given results
of great practical importance. These mathematical models are usually
limited to the processes inside the product, and their solutions are
independent of the processes of heat transfer from the heaters to the
product surface or of vapour transport away from the product surface.

These processes, which are external to the product under freeze drying
but internal in the freeze drying equipment, are indispensable parts of the
total freeze drying process. They are even the most important parts to
master, when it comes to questions of how to design efficient equipment for
freeze drying. This outside view on the freeze drying process has to be
considered carefully.

1.1. The Prefreezing Step

The freezing of clean water normally takes place at a constant temperature of 273 K, ice crystals forming and growing from the liquid water gradually as heat is removed. To produce 1 kg of ice crystals at 273 K from water of 273 K, 334 kJ of heat has to be removed. This simple physical relation is not sufficient to understand the aspects of the prefreezing step of the freeze drying process.

In freeze drying clean water is not considered a food. But water is an important component of all foods, and for the freeze drying process it is the all important product component. This water is normally present as part of aqueous solutions and suspensions, and the freezing of those is connected with phenomena such as freezing point depression, eutectic solutions, subcooling, ice crystal growth and glassy structure. We will first consider liquid foods, all of which lack an internal structure before the prefreezing step.

In the prefreezing step of a liquid food, an internal structure is created in the food material, which may be stabilised during a later sublimation drying and secondary drying. This structure is formed by a network of ice crystals incorporated in a matrix of concentrated food material. In the latter steps of the freeze drying process the ice crystals are sublimed away leaving a network of channels for the water vapour to escape to the product surfaces. The water content of the matrix reaches these channels by diffusion and also escapes to the surfaces. At the end of the freeze drying process the former liquid material is transformed to a porous solid with the concentrated solid matrix as a structural network.

When lowering the temperature of a liquid food, freezing will start at a temperature some degrees below 273 K depending on the strength and composition of the solution. Ice crystals will form leaving a stronger solution with a lower freezing point. The crystallisation will then proceed at an ever decreasing temperature and in an ever more viscous solution. The viscosity is increasing both from the lowering of the temperature and from the increasing dry solids (DS) content. At a certain temperature the remaining solution freezes to a solid mass. This may be the case of the crystallisation of a eutectic solution,[6] but with foods it is as often the case of the formation of a glassy structure known as the vitreous state. This is the normal case with sugar-containing solutions like fruit juices. In these cases, the viscosity is so high that the material gives the impression of a solid material. The water molecules are immobilised, so that the crystallisation of ice can no longer proceed. With appropriate temperature treatment of such glass, further crystallisation may be obtained; known as devitrification.

This effect is studied with techniques known as differential thermal analysis (DTA) and differential scanning microcalometry (DSC).[7]

Sugar solutions are very apt to form glass structures, which give high resistance to vapour diffusion. This may cause difficulties in the later steps of the freeze drying process.

In such cases, other means of influencing the structure of the prefrozen material may be used. One way is to freeze the liquid as a foam, so that gas bubbles are locked into the frozen matrix. This can provide extra possibilities for vapour diffusion. This foam freezing method has been extensively used in freeze drying of concentrated coffee extracts. It also serves the purpose of controlling the ultimate density of the freeze dried product.

Another example of improving the structure is in the production of freeze dried orange juice. When fruit pulp is suspended in the juice, the freeze drying process goes much easier than with a clear juice without any pulp.

When large ice crystals are formed, a structure with larger size channels will result, which will ease the water vapour transport from the product interior to the surface. Larger ice crystals are favoured by a slow freezing process.

In the prefreezing of solid foods, the structural influence is also of importance. Solid food products have a well defined natural structure,[8] but the prefreezing may to some extent transform this structure. Slow freezing favours the formation of large ice crystals, which can penetrate and damage cell membranes. Also ultraquick cryogenic freezing may result in drastic structure break down caused by temperature stresses. These structural changes are normally undesired in quick frozen foods. However, an opened structure will favour the water vapour transport out to the product surfaces in the later steps of the freeze drying process, and it might also favour the rehydration property of the freeze dried product. For freeze dried products a good compromise is often to use semiquick prefreezing.

So far we have considered the influence of the prefreezing step on the microstructure of the product. Of equally great practical importance is the macrostructure of the product, as it is given by the size and shape of the individual product pieces. With the same microstructure, naturally the water vapour will escape more easily to the surfaces of smaller size particles than of larger size particles.

The macrostructures of solid products are usually formed before the prefreezing step in operations such as slicing and dicing. The cutting through of product membranes in such preparatory operations may be decisive for sufficient water vapour passage.

Liquid products are often given their macrostructure after the prefreezing step. When the temperature is kept low enough, the prefrozen product can be milled to the desired granule size and sifted to avoid fines in the end product.

1.2. Sublimation Drying

In the sublimation drying step the two main transport processes are heat transfer and water vapour transfer. But other phenomena are also of importance, for instance chemical reactions in the material and transfer of aromatic volatiles.

1.2.1. Heat Transfer

In the normal vacuum freeze drier heat is transferred to the product surface and on through the product layers to the sublimation front. The driving force of this heat transfer is the temperature difference from the heater surface to the sublimation front. The transport means are the sum of the heat conduction, the heat convection and the heat radiation. The total temperature difference is divided between the two main sections of heat transfer, the first one being from heater surface to product surface, and the other one being from product surface to sublimation front. These two sections represent a series connection of two heat transfer resistances.

The highest product temperature will be found on the product surface, where heat is transferred from the heater surface. To avoid serious quality losses it is important to keep this product temperature below a certain limit, which can be fixed individually for each kind of product.

It is the object of the heater regulation system to control the heat input to the product, so that the permissible product temperature is not exceeded but kept close to this level for process speed and economy.

The heat transfer in the vacuum chamber between the heater surface and the product surface has been the subject of numerous works. The different solutions to this task have given names to different main types of freeze driers. In these cases the sublimation drying takes place in shallow product layers, and the normal form of heater surface is the horizontal heater plate.

The first type of freeze drier was the simple contact freeze drier.[9] Here the product is placed in a shallow layer in a metal tray, which is placed on the horizontal heater plate. The heat transfer from the heater plate to the product surface is mainly by conduction through the metal tray in contact with the heater plate and with parts of the product surface. If contact conditions are so good that an even heat distribution may be obtained, the sublimation drying can be carried out under sufficiently controlled conditions. The process time is relatively long, because the heat transfer is mainly to one side of the product layer only.

In the 1950s the double contact freeze drying system was developed under the name of accelerated freeze drying. By establishing heat transfer to both sides of the product layer, the process time could be drastically reduced.[10] Here again shallow product trays were used. Beneath the product layer in the tray a sheet of expanded metal was placed, and another sheet was placed on top of the product layer. This sandwich was placed between horizontal heater plates, which were pressed together against the interleaved product. Heat was transferred by conduction (contact) from the heater plates through the tray and through the expanded metal sheet to each of the two product sides. The expanded metal fulfilled a double purpose. It provided channels for water vapour escape between the product and the metal surfaces, and it gave an even heat distribution over the product surface. Variation in product layer thickness was compensated for by local impression of the expanded metal into the product surface.

Another development from the simple contact system is the ribbed tray contact system.[11] Here the prefrozen product is filled into an aluminium tray with high integral ribs on the product side. This gives an extended contact surface on the product side.

In contrast to the contact freeze drying systems (which rely on conductive heat transfer through metal from heaters to product surfaces) are the radiation freeze drying systems with main heat transfer between heaters and product surfaces by radiation transfer, without metallic contact.

In the simple radiation freeze drying system[12] product trays with a shallow product layer are interleaved between parallel heater plates with a free space all around the heater plates. Heat is transferred by radiation through this free space from the heater plate above to the product top surface and from the heater plate below to the tray bottom, which is in contact with the product. Also in radiation freeze drying ribbed trays have been used to increase heat transfer to the product.[13]

The objective of the heat transfer system is to transmit a controlled heat flux evenly to the whole heat-receiving surface of the product. The control of the heat flux is effected by controlling the temperature of the heater plates, the even heat distribution is safeguarded by the geometry of the system.

In contact systems the heat transfer is mainly by heat conduction through the metal of the tray and through any vapour gaps between the tray and the heater plate. A vapour gap is an effective heat insulation compared to direct metallic contact, so minor unevenness of heater plates or tray surfaces will give detrimental unevenness of heat distribution. This is a definite weakness of the contact freeze drying system, because it is not so simple to maintain large surfaces of this kind in a geometrical flush condition. The double contact system can better adjust to practical conditions by the pressure put on to the surfaces and by the expanded metal's variable penetration into product surfaces.

Even heat distribution is maintained much more simply in the radiation freeze drying system. The radiation heat transfer between wide parallel surfaces is independent of the distance between the surfaces. This means that normal technical unevenness of trays or heater plates will not at all disturb the heat distribution.

With distances of 10 mm or more the radiation heat transfer is the dominating component. With distances of 1 mm or less the conductive heat transfer through the vapour is the dominating component. In any case the convective heat transfer component is so small that it can be disregarded, because the vacuum vapour has a very low heat carrying capacity.

To get the same heat flux to the product surface in a radiation freeze drying system may require a somewhat higher heater plate temperature than in a contact system. But even in a radiation system moderate heater temperatures ($\simeq 400$ K) are sufficient for all practical purposes, so this actually presents no complication.

The process conditions in the material itself depend on the heat flux to the product surfaces, but it is of no concern whether this heat flux is produced by 'contact' or by 'radiation'. So the heat transfer within the product material may normally be studied by itself and the heat transfer system in the freeze drier disregarded.

The heat transfer within the product material itself will differ according to the material configuration. In blockform material this heat transfer can be regarded as heat conduction, and normally the heat has to pass through already dried layers of material forming a good heat insulator. This heat conduction will often be rate limiting for the process. Mathematically this process is not as simple as it may seem[14] because the heat conductivity of the material will vary as the layers gradually dry out during the secondary drying.

Several methods have been proposed to increase the efficiency of heat conduction into block material. Heat input through frozen layers is theoretically possible[15] by freezing the product on to the tray bottom. In practice, however, it has shown to be next to impossible to obtain evenly controlled processes. Starting from the edges of the block, a dry surface layer will form and work its way successively in along the tray surface and break the thermal contact between the tray metal and the frozen mass of the block.

Spiked plate heaters[16] for metal conduction of heat into the block interior have also proved impractical.

Infrared heat radiation of wavelengths around 1 μm has shown to have considerable penetration effects into various foods.[17] To produce this wavelength a heater temperature of approximately 3000 K is required, which calls for glass capsulated radiators (lamps). This is a definite drawback of the method.

A similar penetrating effect can be obtained with dielectric[18] heating and with microwave heating.[19] This energy form also has the advantage that it concentrates more in the more humid parts of the product than in the dry sections. The technique however is so complicated that so far it has only been used in small test plants.

When operating with a shallow bed of reasonably coarse granules any wavelength of infrared heating radiation will have a good penetrating effect. This is commonly used, as the granular form of freeze dried products is generally popular.

When continuously mixing a bed of granulated product so that new product surfaces serve as heat receivers, the heat flux into the product can be increased drastically.[20]

1.2.2. Water Vapour Transport

At a moderate freeze drying vacuum of 1 torr each gramme of ice when sublimated produces approximately 1 m^3 of water vapour and at 0·5 torr even 2 m^3 may be formed. Being produced at the sublimation front all this water vapour has to be transported first to the product surface and then away from the product surface, otherwise the vapour pressure will rise and the temperature at the sublimation front will also rise so that melting of the frozen product will occur. This vapour transport must take place under limited pressure differential, or likewise melting may occur, which may first be recognised as collapse of the regular, porous structure of the freeze dried product.

For true freeze drying to take place, the vapour pressure at the sublimation front driving the vapour out of the product should never be higher than the saturation pressure of water at a temperature where the product is still hard frozen. Incipient melting at the sublimation front should be avoided.

The freeze drying process can also be carried out under normal atmospheric pressure.[21] However, under those conditions the water vapour transport inside the product as a diffusion process is extremely slow, and an economic process solution has not yet been found. All freeze drying plants therefore operate as vacuum freeze driers. And even so the vapour transport out of the product may often be the rate limiting factor of the sublimation process.

In industrial freeze drying, mechanical pumping of the vast vapour volumes produced is not economical. Normally the water vapour is removed by condensation on a cold surface, known as the ice condenser or vapour trap.

The flow of water vapour from the sublimation front will flush non-condensable gases along to the vapour trap. A vacuum pumping system connected to the vapour trap removes the non-condensable gases to prevent a pressure build up. Partial pressures of gases other than water vapour are, therefore, so low in vacuum freeze driers that water vapour transport takes place as mass flow.

The flow regime of water vapour in vacuum freeze drying depends on two dimensionless numbers, the Reynolds number, Re, and the Knudsen number, Kn. $Kn = (Xm)/l$, where Xm is the mean free path of the vapour molecules and l is a characteristic dimension of the vapour flow channel.

If Kn is much larger than 1, the vapour molecules bounce much more frequently against the channel walls than against each other. This situation will be normal in the pores of the product, and this flow regime is known as molecular flow.[22]

When Kn is much smaller than 1, the vapour molecules bounce much more frequently against each other than against the channel walls and we have the conditions of ordinary fluid flow. Then the fluid flow regime—a turbulent flow or a laminar flow—will depend on Re. $Re = (Cl)/v$, where C is the mean vapour flow velocity, l is a characteristic dimension of the vapour flow channel and v is the kinematic viscosity of the vapour.

The kinematic viscosities of vapours are inversely proportional to the vapour pressure. Vacuum vapours consequently have relatively high kinematic viscosities, so the flow regime of the vapours outside the product is usually laminar flow.

When the flow path is known to its exact dimensions, the water vapour flow and the corresponding pressure drop can be calculated according to the physical laws for the kind of flow regimes in question.

In an idealised model product with well defined pore dimensions and heat conductivity, the whole sublimation process can be calculated exactly as simultaneous heat transfer from heater plates via product surface to the sublimation front and water vapour transfer from the sublimation front via the product surface to the vapour trap.[23]

In real life however the heat conductivity of the product varies during the process, and so do the dimensions of the product pores, which will be evident as product shrinkage is observed. Therefore laboratory test runs of the freeze drying process with the products in question will always be necessary to get exact and reliable process information.

1.3. Secondary Drying

The sublimation drying of the ice crystals leaves a system of pores in a matrix of product. The rest of the water content of the product has not been frozen out as ice crystals and is still distributed within the matrix material. The bulk of this moisture will be removed during the secondary drying step. This moisture may partly exist in the vitreous state, partly as hygroscopically bound water.   ·

The secondary drying step is very similar to the latter steps of any drying process,[24] when water is no longer mobile in the product in the liquid form and a temperature rise is necessary to release the hygroscopically bound water. Foods normally in the secondary drying can tolerate the sufficient rise in temperature over the sublimation temperature to achieve this. For some pharmaceuticals, however, a higher vacuum is necessary for the secondary drying step. For foods the normal operation is that sublimation drying and secondary drying take place simultaneously in different parts of the product and naturally in the same piece of equipment.

Theoretically the best product preservation is achieved when the residual moisture content is brought down to a monomolecular layer of absorbed water according to the BET equation.[25] For practical purposes 2 % residual moisture content may serve as a guideline.

The secondary drying step is also like the sublimation drying step; a combined process of heat transfer and of water vapour transfer. Because of the complexity of this process step calculations are of very limited value. Real knowledge of this process step can be gained by laboratory testing methods.

The practical aims of these tests are to define the end point of the freeze drying process, so that the specified residual moisture content is obtained. Three different methods have been used, the pressure rise method, the temperature approach method and the weight method. In the pressure rise method[26] the vacuum cabinet is shut off from the vapour trap, and the subsequent rise in the cabinet pressure is observed. At specified process conditions the rate of this pressure rise is characteristic of the residual moisture content. This rate naturally depends also on product quantity and vacuum cabinet size, and this makes it difficult to use that method in practice.

In the temperature approach method product temperatures are observed during the conclusion of the freeze drying operation. Product temperatures tend to approach the heater plates' temperature as the product dries out. The temperature difference between heater plates and product will be characteristic of the moisture content. The measurement of exact product temperatures during the process is a delicate task, but when standardised measuring techniques are used, the temperature difference readings will serve the purpose nicely. As it is easy to standardise temperature measurements, this method has found wide application.

When using the weight method a product sample is weighed continuously or at regular intervals during the process. The rate of weight reduction per gramme of product is then a measure of the residual moisture content of the sample. This method can well be used in laboratory testing equipment. In industrial operations, however, the weighing of the same sample is not always practicable. In laboratory testing this method is often used together with the temperature approach method. The correct temperature approach can be established quickly and can be used for measurements in industrial operations.

In the freezing step the product volume increases because ice crystals have about 10% higher volume than liquid water. In the sublimation step the product volume will remain fairly constant. Even when the ice crystals have sublimed away, the remaining product matrix is rigid enough to keep its shape at the sublimation temperature.

However, with the temperature rise in the secondary drying step, the rigidity of the matrix diminishes, so a volume shrinkage will often occur, although not as obviously as in hot air drying of products. The extent of this shrinkage will depend largely on product characteristics but also on process conditions. Many products can be freeze dried without significant shrinkage, but in some cases as much as 30% volume reduction may be observed.

1.4. Collapse

Most products can be readily freeze dried resulting in the characteristic, porous freeze dried structure. But with some products difficulties in conducting a true freeze drying process arise. The freeze dried structure tends to collapse, and related problems like puffing or local melting may occur. Collapse[27] means that the matrix structure sags during the process, so that pores diminish and eventually get closed giving rise to puffing and a changed structure in the end product.

Products with a high tendency to form glassy and amorphous structures when prefrozen are more liable to collapse. Collapse is generally avoided if the temperature level can be kept low enough. The temperature level at the sublimation front is a result of the cabinet pressure plus the pressure differential of the vapour flow from the sublimation front to the product surface. The sublimation front temperature can be lowered by reducing the cabinet pressure or by reducing the vapour flux, which may be effected by reducing the heater temperature. The pressure differential will also be lower in smaller product pieces than in larger ones. The product temperature in any point between the sublimation front and the product surface is an intermediate between the sublimation front temperature and the surface temperature. The surface temperature will also be lower with a reduction of the heater temperature.

If collapse occurs, the first remedy is therefore to reduce the heater temperature and/or the pressure of the vacuum chamber or to operate with smaller product particles. When freedom in product formulation exists, the possibility of adding or reducing the contents of specific components to raise the collapse temperature may be the most economic solution to the problem. For instance orange juices containing the fruit pulp freeze dry much easier without collapse than do filtered juices.

Experiments[28] with 25 % fructose solutions give collapse temperatures of 229 K without additives, 232 K with 1 % gelatin, 245 K with 2 % gelatin and 239 K with 1 % pectin.

Collapse is a sudden and irreversible phenomenon, because when it starts closing the product pores this further aggravates the condition.

1.5. Batch Freeze Drying

In batch freeze drying a whole charge of prefrozen products is prepared in the trays. Awaiting the sublimation drying step this charge has to be kept under low temperature.

The full charge is then brought quickly into the vacuum cabinet with each tray in the correct position in relation to the heaters. The cabinet is closed and evacuated down to the correct freeze drying pressure. When the charging and evacuating operations are effected fast enough, there is no need to place the vacuum cabinet or part of it in a refrigerated room. The heaters in the cabinet should be chilled to ambient temperature before the charging operation. If parts of the product have picked up some heat during the charging operation, the product temperature will, shortly after the evacuation, be brought back to normal by slight evaporation from warmer zones.

The heating is then turned on and the heater temperature is controlled to a constant level, for instance 400 K, over the entire heater surface. This heater temperature is kept constant over a certain period of the drying cycle. Then with the progress of the sublimation and secondary drying of the product, the heater temperature is gradually reduced to an end point temperature of, for instance 325 K, where it may be kept constant again for a period of time.

This temperature–time programme has to be decided for each individual product type and each tray loading by laboratory tests with the products. Normally tray loadings are chosen giving freeze drying cycle times of 6–8 h.

When the freeze drying process is finished the cabinet vacuum is broken by opening, allowing the entrance of atmospheric air, or in special cases dry nitrogen. When the cabinet is opened the product charge is withdrawn for packaging.

In batch freeze drying the total cycle time is the sum of the freeze drying process time and the cabinet operation times. So the cabinet operation times should be as short as possible. This means that the product must be brought in and out without delay, that evacuation and vacuum breaking must be carried out quickly, that plate cooling and plate heating are effected rapidly and that deicing of the vapour trap can take place during the freeze drying operation. In an efficient plant the whole cabinet operation time of the cycle should be down to 15 min.

1.6. Continuous Freeze Drying

In continuous freeze driers the vacuum is kept in the freeze drying cabinet throughout the operation. At short time intervals small portions of the prefrozen material are sluiced into the cabinet through a vacuum lock. Each product portion is carried through the cabinet past the heater plate system and ultimately sluiced out of the chamber as finished product for final packaging.

Here the heaters are kept at constant temperature in each of several zones along the path of the product through the cabinet. In this way each portion of the product runs through the same process as in batch freeze drying. The temperature programme for continuous freeze drying is established the same way as in batch freeze drying by laboratory tests with the products in question.

2. FREEZE DRYING EQUIPMENT

When freeze drying foods, separate equipment, usually a freezer, is used for the prefreezing step. The sublimation drying step and the secondary drying step of the process are carried out simultaneously in a vacuum cabinet, known as the freeze drier.

2.1. Freezers

For the prefreezing process normally the same types of freezers are used as in the quick frozen foods industry. The choice of freezer type depends largely on the type of product but to some extent also on the type of freeze drier used. For freeze driers with straight trays, the products may often be prefrozen in the same trays carried through a conventional air blast freezing tunnel.[29] Particularly for batch freeze driers, this may give a simple and practical installation.

When a free-flowing product of relatively small but uniform particle size is produced, a freezer giving a free-flowing prefrozen product, e.g. IQF-freezer, is favourable because such free-flowing prefrozen material can easily be charged into freeze drier trays.[30] A typical freezer of this kind is the fluidised bed freezer. IQF-freezing is most suitable for continuous freeze drying.

With liquid raw material a free-flowing prefrozen granulate is often preferred. Prefreezing in a thin layer on a rotating drum has shown great advantages.[31] The prefrozen slices that are released from the freezing drum are then dropped into a mill situated in a cold room to produce a granulate of the desired particle size.

Prefreezing of liquids in blocks or lumps followed by crushing in mills in a cold room has also been used, but the technological difficulties increase sharply with the block size.

Also prefreezing of liquid products by dripping into a cold bath of liquid nitrogen or other refrigerant has been proposed.

2.2. Batch Freeze Driers

The various batch freeze driers are mainly characterised by the method of heat transfer between heater plates and product trays and by the shape of the product trays and by the design and arrangement of the ice condenser system.[32]

Taking a look at a popular type of batch freeze drier (Figs 3 and 4) these main characteristics will be discussed.

In Fig. 4 heat transfer between the heater plates 4 and the product 7 is by radiation from the heater plate 4 to the bottom of the straight product tray 5 and from the heater plate 6 to the top layers of product 7.

An even surface temperature of the heater plates safeguards in the simplest possible way an even heat distribution to the main product

FIG. 3.   Product charging in a radiation freeze drier.

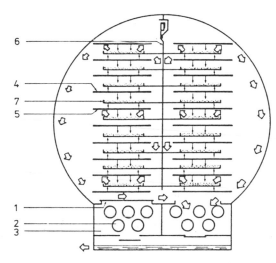

FIG. 4.   Operation principle of a radiation freeze drier with automatic ice condenser.

surfaces. Likewise this arrangement safeguards the easy escape of the water vapour from the product bed. The straight tray is also most versatile with regard to product type and product shape. It is important that the product is distributed in an even layer in the product tray.

An alternative heating system is the contact heating system with ribbed trays. The practical difficulties with this system to achieve even heat distribution has been mentioned earlier. Here the product layer thickness is fixed as the distance between the tray ribs. However, some product forms will not be suitable for such fixed layer thickness.

The purpose of the ribbed tray is to achieve a higher heat flux. This higher heat flux is linked with a proportional water vapour flux out of the product bed. In freeze drying this vapour flux has very definite limitations, otherwise product particles may be thrown out of the bed.[33] To channel the water vapour away from the product, more complex ribbed trays have been introduced.[34]

Heat transfer systems with moving product in direct contact with heater plates have been advocated.[35] This may give extremely high heat fluxes but the water vapour fluxes inevitably carry material away, which has to be collected and processed separately.[36] This system is also problematic from the hygiene point of view.

The ice condenser 2 of the batch freeze drier shown is built into the vacuum cabinet with good flow conditions for the water vapour. The deicing procedure operates automatically, so that only a thin ice layer is allowed to build up before being melted off. The condensing surface therefore operates at full capacity at any time.

Other freeze driers operate with ice condensers in separate vessels connected through pipe lines with valves, normally designed for deicing at longer intervals, usually only between batches. This gives rise to pressure drops in the vapour path and to temperature differentials through heavy ice layers. This necessitates lower refrigerant temperatures for satisfactory operation.

The major part of the whole energy consumption on freeze drying is spent on refrigerating the ice condenser. Thus the thermal efficiency of the ice condenser is vital to an economic operation of the freeze drying production.[31]

2.3. Continuous Freeze Driers
Figure 5 shows an example of a continuous freeze drier that has found general acceptance.[37] It is shown opened at the end to show some of the interior, but in operation it has to be closed for the vacuum.

FIG. 5. A continuous radiation freeze drier for products in straight trays.

The product is carried as a shallow stationary bed in straight trays. The trays enter the long vacuum cabinet one by one through a narrow vacuum lock seen to the right in front. The vacuum lock can be actively sealed off from the atmosphere as well as from the vacuum cabinet. After a tray is brought in, the lock is sealed and evacuated by a separate vacuum pump down to freeze drying pressure. Then the connection to the vacuum cabinet is opened and the tray transferred. In the entrance end of the cabinet the trays are successively collected 15 to a stack, which in one movement is pushed one step forward through the cabinet, each tray in a pair of guide rails.

Each pair of rails guides a continuous line of trays in correct radiation position between the stationary heater plates through the whole length of the cabinet. In the exit end of the cabinet a stack of 15 trays is pushed out when the stack is pushed forward in the entrance end. The trays of the exit stack have each moved down between the whole line of heater plates during 6 h, so by this time the freeze drying process is finished. This stack of trays is transferred one by one out through an exit vacuum lock of the same design as the entrance vacuum lock, which is also operated by means of a separate vacuum pump. The empty exit lock must be evacuated down to the freeze drying pressure, before it is opened to the vacuum chamber for acceptance of a tray with finished product.

Another continuous freeze drier operates with vacuum locks for a whole rackcar with ribbed product trays at a time. In this case the vacuum lock forms one extra length section of the vacuum cabinet. This gives a relatively large vacuum lock to evacuate quickly, which takes a relatively large vacuum pump.

Two different types of continuous freeze driers with the product particles moving in direct contact with heater plates have been introduced. In one type the product is dropped through a vacuum lock down on to the top plate in a stack of circular heating plates.[38] Rotating scrapers work the product successively outward on the top plate, until it falls over the plate rim down on to the next plate. Here the product is worked similarly inward on the plate until it falls over the inner plate rim down to the next plate and so on past the whole stack of plates. This gives a good mixing and an excellent heat transfer, but it also gives product abrasion and much product fines flowing out with the water vapour stream.

In another type the product is moved by vibration of the generally horizontal long heater plates. In this case too the product abrasion gives serious complications.[39]

Continuous freeze driers are generally more economic than batch freeze driers, they have lower energy consumption, need less operating staff and they may have a higher degree of operation reliability, mainly because they can eliminate numerous human errors.

3. THE FREEZE DRIED PRODUCTS

The freeze drying process is an expensive preservation process, more so than hot air drying or quick freezing. The quality of freeze dried foods are generally found to be closer to quick frozen foods than to air dried products. Freeze drying therefore has its place where quick frozen or air dried foods cannot be used.

3.1. Quality Factors
The nutritive and organoleptic quality of freeze dried foods are comparable with quick frozen foods, because the natural qualities of the products such as vitamins, taste, flavour and appearance have been preserved basically in the same way.

When compared to quick frozen foods, the freeze dried products have the advantages that they have reduced weight and that they keep their quality at normal ambient temperature. This is, for instance, the reason why quick frozen coffee extract is no competitor for freeze dried coffee. The product must keep its free flowing character to be able to be measured spoon-wise for brewing several times. The freeze dried product can conveniently be used as an ingredient in dry product mixes, where the quick frozen counterpart is not suitable.

When compared to hot air dried products, the main quality advantage of freeze dried foods are appearance, flavour retention and rehydration ability. Generally, the fresh colour of the product is maintained and the shape is only slightly changed, so that the product can easily be recognised for what it is. In dry food mixes, therefore, often some of the most characteristic components will be freeze dried.

The rehydration ability of freeze dried foods gives special value in instant foods, when a hot dish can be presented in seconds only by addition to the preserved food of hot water. This simplifies automatic serving of hot dishes.

However, the most recognised feature of the freeze drying process is the amazing quality of aroma retention. Aroma retentions, as measured by gas chromatography, of 80–100% are often achieved.[40] This result is remarkable when one considers that the aroma components are far more volatile than water. So in wet drying the aroma components will leave the product more readily than the water. In freeze drying the aroma components are distributed, within the matrix structure. The clean water of the ice crystals can escape without entraining the volatiles. The water fraction in the matrix and the volatiles can escape through diffusion through matrix material to pore system. Here the volatile aroma components are trapped because they diffuse so much slower through the matrix material than the water vapour.[41] And as the matrix is dried out, the diffusion is successively slowed down, so in the finished product the aroma components are practically locked in.

Quick frozen foods generally should be protected by low temperature storage as well as by packaging. The packaging material should provide a hygienic environment, and it should be tight enough to give protection against desiccation. Freeze dried foods need stronger protection from the packaging material, because they are kept at ambient temperature. In all cases a safe protection against water vapour entering is necessary. Also very often protection of the freeze dried product against the oxygen of the air is required. Some products may also be harmed by light. Glass and three-ply laminated foil packages of polyethylene, aluminium, and polyester are the favoured packaging materials. The products most sensitive to oxygen should be protected by vacuum packaging or flushed with nitrogen or other protective gas.

3.2. Economic factors

Freeze drying has generally been recognised as a very expensive preservation method, but still in most cases raw material costs, packaging costs and distribution costs are the dominating ones. So in many cases the special product advantages that freeze drying will give outweigh the extra production costs incurred.

The 'instant' factor may be a very important economic factor, when it comes to customer capacity of restaurants or hot dish automats. It is also attractive for simplified housekeeping.

The low weight and the stability at ambient temperatures make freeze dried foods competitive when the distribution system is not a streamlined one. Distant camps and expeditions are extreme examples of this kind.

In cases where other preservation methods incur important product losses, freeze drying is also economic. A typical example is spicy herbs which are used in small quantities at intervals, so that the keeping quality of the freeze dried product gives full value.

Experience has shown that seemingly slight improvements in convenience, aroma, instant quality or keepability have given an economic basis for freeze dried products, so there are still a number of new possibilities for this interesting preservation process.

REFERENCES

1. REY, L., In *Researches and Development in Freeze-drying*, Hermann, Paris, 1964.
2. LORENTZEN, J., *Kulde*, 1967, **21**, 68.
3. LORENTZEN, G., 'Atlas Freeze Drying Symposium', Copenhagen, 1970.
4. KESSLER, H. G., *Kältetechnik*, 1962, **14**, 174.
5. LORENTZEN, J., 'Atlas Freeze Drying Symposium', Copenhagen, 1970.
6. KUPRIANOFF, J., In *Researches and Development in Freeze-drying*, Hermann, Paris, 1964.
7. MALTINI, E., In *Freeze Drying and Advanced Food Technology*, Academic Press, London, 1975.
8. BENGTSSON, N. E., In *Freeze Drying and Advanced Food Technology*, Academic Press, London, 1975.
9. ROWE, T. W. G., *N.Y. Acad. Sci. Ann.*, 1960, **85**(2), 641.
10. FORREST, J. C., In *Freeze-drying of Foodstuffs*, Columbine Press, Manchester, 1963.
11. HAMILTON, W. E., *et al.*, US patent 3.247602, 1966.
12. KAN, B., *et al.*, *Fd Technol.*, 1968, **22**, 67.
13. NERGE, W., *et al.*, US patent 3.270433, 1966.

14. MINK, W. H., *et al.*, In *Freeze-drying of Foods*, Nat. Acad Sci., Washington DC, 1962.
15. LAMBERT, J. B., *et al.*, In *Freeze-drying of Foods*, Nat. Acad Sci., Washington DC, 1962.
16. BALLANTYNE, R. M., *et al.*, *Fd Technol.*, 1958, **12**, 398.
17. GINSBURG, A. S., *Infrarottechnik und Lebensmittelproduktion*, Fachbuchverlag, Leipzig, 1973.
18. LEATHERMAN, A. E., *et al.*, In *Freeze-drying of Foods*, Nat. Acad. Sci., Washington DC, 1962.
19. DECAREAU, R. V., In *Freeze-drying of Foods*, Nat. Acad. Sci., Washington DC, 1962.
20. KESSLER, H. G., In *Freeze Drying and Advanced Food Technology*, Academic Press, London, 1975.
21. GUOIGO, E. I., *et al.*, *Bull IIR Annexe* 1974–3, 137.
22. VAN ATTA, C. M., *Vacuum Science and Engineering* McGraw-Hill, New York, 1965.
23. TRIFONOVA, L. I., *et al.*, *Proc. XIV Intern. Congress Refr.*, Moscow, Vol. 3, 1975, p. 270.
24. KRISCHER, O., *Die wissenschaftlichen Grundlagen der Trocknungstechnik*, Springer, Berlin, 1963.
25. SALWIN, H., In *Freeze-drying of Foods*, Nat. Acad. Sci., Washington DC, 1962.
26. KAN, B., In *Freeze-drying of Foods*, Nat. Acad. Sci., Washington DC, 1962.
27. MACKENZIE, A. P., In *Freeze Drying and Advanced Food Technology*, Academic Press, London, 1975.
28. KING, C. J., In *Freeze Drying and Advanced Food Technology*, Academic Press, London, 1975.
29. HEWITT, M. R., In *Freeze Drying and Advanced Food Technology*, Academic Press, London 1975.
30. PERSSON, P. O., In *Freeze Drying and Advanced Food Technology*, Academic Press, London, 1975.
31. LORENTZEN, J., *Proc. XIV Intern. Congress Refr.*, Moscow, Vol. 3, 1975, p. 140.
32. LORENTZEN, J., In *Freeze Drying and Advanced Food Technology*, Academic Press, London, 1975.
33. DAUVOIS, P., *Revue général du Froid*, 1971, 827.
34. EILENBERG, H., *et al.*, US patent 3.401468, 1968.
35. KESSLER, H. G., US patent 3.460269, 1969.
36. SCHIMPFLE, J., US patent 3.936952, 1976.
37. LORENTZEN, J., *Chemistry and Industry*, 1979, 465.
38. DALGLEISH, J. M., In *Freeze-drying of Food Stuffs*, Columbine Press, Manchester, 1963.
39. ROTHMAYR, W., US patent 3.465452, 1969.
40. FLINK, J., In *Freeze Drying and Advanced Food Technology*, Academic Press, London, 1975.
41. THIJSSEN, H. A. C., In *Freeze Drying and Advanced Food Technology*, Academic Press, London, 1975.

Chapter 6

EXTRUSION PROCESSING—A STUDY IN BASIC PHENOMENA AND APPLICATION OF SYSTEMS ANALYSIS

Juhani Olkku

Technical Research Centre of Finland,
Espoo, Finland

SUMMARY

Extrusion cooking is an increasingly important processing technique in modern food and feed industries. Extrusion cookers are considered to be general purpose, thermodynamically polytropic screw reactors, where 'reaction analogous' basic changes in biopolymer aggregate structures and chemical reactions proper take place. The biopolymer aggregate structure changes are discussed within the physical conditions encountered in extrusion cookers. The changes begin with an irreversible denaturation of nature's biopolymer aggregates. This is a necessary prerequisite for the plastification through new aggregate and texture formation during the process. Denaturation and new aggregate formation seem to be the predominant mechanisms up to mass temperatures around 170°C, after which the biopolymers seem to start to lose their aggregate forming ability due to the affects of temperature and shear. This heuristic and hypothetical theory gives a basis for the prediction of the effects of common ingredients such as salt, sugar and monoglycerides when applied in recipes to be extrusion cooked.

Stage-wise, descriptive systems analysis is performed on the functional sections of a cooking extruder followed by a discussion of mass states and the factors affecting the mass stages in each of the functional sectors. Material–machine interactions are strongly emphasised.

Single-screw and twin-screw extruders are compared descriptively, and differences in their functional performance are commented on. Applications

177

*and the role of mathematical modelling are discussed. Two strategies are
applied. These are the empirical black box approach and the approach based
on fundamental physical principles. Both have their merits and shortcomings
when modelling a process like extrusion cooking. Simulation and
mathematical modelling are considered to be the key to the solutions through
which extrusion will be transformed from art to science.*

1. INTRODUCTION

Extrusion cooking is an increasingly widespread processing technique in
food and feed industries. New applications are constantly appearing for
this continuous, space and energy efficient technology. Practical industrial
applications are in this case ahead of research findings and theory
development. The advantages of the technique are, however, large enough
so that the art of extrusion daily gains new practitioners.

Harper[19] has recently published a review on food extrusion. General,
descriptive articles on the technology have been written by Smith,[50] Smith
and Ben-Gera,[51] Clark,[11] Horn and Bronikowski[23] and Williams *et
al.*[55,56] The classical article on the modelling aspects in food extrusion is by
Rossen and Miller.[48] This art is also presented in a series of articles in *J.
Food Process Eng.*[9,10,26,39,46] where problems arising in mathematical
modelling of extrusion have also been discussed.

This chapter is not intended to be a review, or a general description of
extrusion cooking, nor does it concentrate on the general aspects of
economics and lines of equipment used when applying this technology in
practice. Instead it discusses the basic phenomena (denaturation and
plastification of biopolymers, such as starch), which take place in HTST
cooking extrusion in thermodynamically polytropic machines that are
most frequently used in food and feed industries. The central mechanical
processing equipment, the extrusion cooker, is discussed as a general screw
reactor which creates the physical environment where natural biopolymers,
such as starch, proteins and cellulose, change their physical properties in a
'reaction analogous' manner, and chemical reactions take place.

In Section 3 a systems analysis approach is used in a sequential analysis of
an extruder. It involves considering functional segments of the extruder and
mass–machine interactions at various states of the mass when venturing
through an extruder. This is followed by a short comparison of twin- and
single-screw machines, and a short commentary on mathematical
modelling.

Most of the text is descriptive with short notes on the basic heuristic principle, which are followed by a discussion on the application of the presented principles on cooking extrusion.

2. DISCUSSION ON THE BASIC CHANGES IN BIOPOLYMER AGGREGATES AND ON THE CHEMICAL REACTIONS TAKING PLACE IN EXTRUSION COOKING CONDITIONS

2.1. General

The extrusion cooker is a high temperature, short time, medium to high pressure, medium to high shear screw reactor for biopolymer modification and conversion. This approach presupposes knowledge about biopolymer reaction kinetics and mathematical models that can be applied to the design and optimisation of the reactor. No comprehensive theory or models on the reactions and changes in biopolymer structures, however, exists at present for the physical and chemical conditions encountered in extrusion cooking. This is partly due to the fact that most pasting and gelatinisation studies have been performed with much higher water contents than normally encountered and at atmospheric pressure, while the water contents in extrusion range from 5 to 40% (w/w) and pressures range from 0·2 to 50 MPa. Attempts to model mathematically the basic changes in biopolymer properties have been published.[28,43,46,52] Remsen and Clark[46] have considered the basic phenomena taking place in extrusion cooking. The heuristic phenomena level approach, which is given below, is a further development of this kind of thinking.

2.2. General Heuristic Phenomena Model for Extrusion Cooking

In the opinion of the author, some kind of theoretical model is essential, if the changes that take place in a cooking extruder are to be understood and mastered. Equation (1) gives a heuristic model for the changes in material states or properties in thermal processing in an extruder.

$$y = f(RM, I, MC; \tilde{P}, \tilde{T}, \tilde{R}, \tilde{S}) \tag{1}$$

where y is a dependent variable such as mass viscosity in a given point in the extruder, mass pressure or temperature before the die, water absorption or solubility index of the product after extrusion, product density, cold water consistency of ground, water suspended extrudate, etc. It can also be a combination of several such product quality indicators, which amounts to a multi-dimensional quality vector $\mathbf{Y}$. RM is the raw material composition of the processed recipe such as wheat flour, soy protein, etc. I is the ingredients added such as monoglycerides, sugar, salt, etc. MC is the moisture content in the extruder. It is separated from other recipe components due to its prime importance on the basic phenomena. $\tilde{P}$ is the pressure profile in the extruder from feed point to the die exit. $\tilde{T}$ is the temperature profile in the extruder from feed point to the die exit. $\tilde{R}$ is the residence time profile in the extruder from feed point to the die exit. $\tilde{S}$ is the shear profile in the extruder from the feed point to the die exit. f is some mathematical function.

The state of the mass in the extruder and in the ready processed stage, as the dependent variables indicate, is thus a result of the temperature and shear action under pressure while the mass resides in the extruder. The temperature is a resultant of added or removed heat by heat exchanging and by mechanical energy dissipation in the extruder. In some modern, well designed plastics extruders the temperature profiles are regulated within $\pm 0.5\,^{\circ}\mathrm{C}$ around the desired values. Such accuracy is but a dream in most food extruders today.

2.3. Basic Changes in Starch During Extrusion Cooking

The critical factor in food and feed thermal processing is the change in biopolymer structures of nature's products. Let us build a hypothetical model for changes in starch as it is subjected to the $(\tilde{P}, \tilde{T}, \tilde{R}, \tilde{S})$ profile actions in extrusion. We can here use a reaction analogy.

Starch is a biopolymer of α-D glucose units (primary structure), which are α-1,4 bonded in amylose, and α-1,4 and α-1,6 bonded in amylopectin (secondary structure). Amyloses at least are thought to have a helical tertiary structure. Aggregation of amylose and possibly some parts of amylopectin molecules by hydrogen bonding and/or possibly by the action of other bonding material yields a partly crystalline quaternary structure. Native, dry starch is a powder, which in extrusion is converted to a coherent and, in the cool dry state, continuous mass.

If only changes in the physical structures of the aggregates take place, the phenomenon is not a proper reaction; it is a thermal operation. This is why it is described as being 'a reaction analogous change in biopolymer aggregate structures'. Biopolymers are always processed in the presence of water. Thus water is considered to be the second 'reactant'.

When a starch–water slurry is heated, starch first reversibly absorbs water until the so-called gelatinisation temperature is reached. This temperature is connected with abrupt changes in the physical properties of starch; this is due to the changes that take place at least in the quaternary structure of starch crystalline aggregates. Donovan[15] has called this change in nature's biopolymer aggregates 'denaturation'. Denaturation in this chapter denotes the initial, irreversible change in starch and other biopolymer aggregate structures. Water, the second 'reactant', is at least an energy transfer agent in this process, and evidently, at the initial denaturation temperature, water has reached high enough 'reactivity' so that hydrogen and other possible bonds that hold the natural aggregate structure together are at least partly ruptured. Although other molecules than starch and water may be involved,[17,43] we can simplify the inspection of the basic phenomena by considering a starch–water system.

Denaturation of biopolymers is a necessary prerequisite for the dough or melt development and texturisation in extrusion. Mercier and co-workers[35-37] have demonstrated by X-ray techniques that changes in crystalline state occur at this stage. Once denaturation has started, we can assume that it follows first-order kinetics, and that the rate constants depend on temperature according to the Arrhenius equation. The limiting factor in the 'reaction' is the reaction rate below 110°C and water diffusion rate at temperatures above 110°C.[52]

We can assume that starch continuously forms new aggregate structures, which in turn are ruptured by the action of heat and shear as processing continues. In extrusion cooking this denaturation–plastification takes place under pressure and at elevated temperatures. Thus high water 'reactivities' are reached even if the moisture content is relatively low. The effective mixing and shear actions overcome the limiting factor of water diffusion into the denaturing, plastifying mass. The ground, extruded, expanded starchy products show a glass- or plastic-like appearance under the light microscope. The mass looks as if it were amorphous, but Mercier and co-workers[35-37] have shown that with some starches there are crystalline structures, which clearly deviate from the structure of natural starch crystalline aggregates.

One way to observe the progress of denaturation and subsequent plastification is to investigate the pasting behaviour of ground, water suspended starch or starchy material, and to compare it with scanning electron micrographs (SEM), and some other indicators such as the water solubility index (WSI) and water absorption index (WAI). The series of experiments, (discussed below), was performed in order to demonstrate the aggregate level changes.

For the experiments wheat flour was used; it contains proteins, lipids, glycoproteins and other constituents that can be involved in the aggregate structure formation. However, it is the aggregate structures that are

TABLE 1
PROCESSING DATA IN A SERIES OF EXPERIMENTS USED TO DEMONSTRATE WAI AND CONSISTENCY MAXIMUM EXISTENCE IN A TWIN-SCREW EXTRUSION COOKING OF WHEAT FLOUR

Sample	Processing variable levels[a]			Observed process data		WAI^e $g\,H_2O/g\,d.s.$	Functional properties data[a]			
	Screw[b]	Feed rate (g/min)	Screw speed (rpm)	Mass temperature[c] (°C)	Current[d] (A)		WSI^e % d.s.	Consistency[f] Initial (mm)	Peak (mm)	Cooling (mm)
RM	—	—	—	—	—	0·90	7·5	2	40	35
RM + 0·1% Hg$_2$Cl	—	—	—	—	—	—	7·5	—	200	230
6	C	200	100	142	8·5	2·82	8·1	12	163	158
7	C	200	125	141	8·0	3·12	8·3	17	160	153
8	C	200	150	143	8·0	3·52	9·5	30	150	150
9	C	200	175	146	9·0	3·81	11·2	46	127	121
10	C	200	200	148	9·0	3·95	12·1	52	119	110
11	R	600	100	168	21	4·34	10·0	82	107	102
12	R	400	100	172	16	4·70	13·5	115	96	67
5	R	200	125	168	16	4·58	20·5	100	39	36
4	R	200	150	178	14	4·06	31·7	74	20	22
3	R	200	175	185	13	3·13	44·2	33	7	9
2	R	200	200	187	11·5	1·90	61·5	19	4	7
1	R	200	200	191	10	1·44	66·2	15	5	7

[a] Finnish wheat flour of medium coarseness (Vasa Mills Ltd.) extruded in a Creusot-Loire BC-45 twin-screw, intermeshing flight, co-rotating machine. Die parameters: diameter 5 mm, capillary length 29 mm. Distance between screw end and die insert 2 mm. Flour moisture content was 12% (w.b.) and feed moisture content 15% (w.b.). Raw material had a 480 BU maximum in amylogramme and a falling point number of 294. Extrudates for analysis had been ground in a Wiley mill through 1 and 0·8 mm sieves.

[b] Screw composition: C indicates compression screw elements only. R indicates that the last 50 mm screw section before the die has been counter pitched.

[c] Observed mass pressures before the die. The barrel set temperature was 150°C in all experimental runs.

[d] Current drawn by the main motor. For a given screw combination and raw material it reflects well the mass pressure before the die.

[e] WAI and WSI are the water absorption index and water solubility index measured by the method of Anderson et al.[3]

[f] Consistency in pasting experiment in a Haake Rotovisco rotational viscometer using a winged stirrer described in ref. 57. The dry matter was 6·23 g in 45 ml of water. Suspensions were heated from the initial temperature of 30°C by a 2°C temperature rise per minute to 95°C, held there for 5 min and cooled with the same temperature slope (as during heating) to 50°C. The given values are relative in that they are read from observed curves and given in curve heights.

important, and the author considers the basic phenomena inspection to be relevant to this series of experiments.†

Table 1 shows the experimental data. The cooking extruder, which was used in the experiments, was a twin-screw, intermeshing flight, co-rotating extruder (Creusot-Loire BC-45) (see Fig. 14 in Section 4).

Initial consistency of ground, water suspended extrudates and WAI and WSI for the experimental series as a function of increasing 'hardness of treatment' are given in Fig. 1.

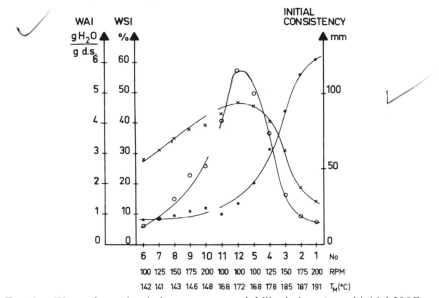

FIG. 1.   Water absorption index, ×, water solubility index, ●, and initial 30 °C consistency, ○, for experimental points given in Table 1. Screw speed and observed mass temperatures are given for each data point.

There is a maximum in WAI and initial consistency values at around a mass temperature of 170 °C. This maximum also exists for rye, barley and oat. The WSI shows a continuous increase that accelerates when the WAI and initial consistency peaks have been reached. Anderson et al.,[3] Conway,[13,14] Lawton et al.,[27] and Mercier and Feillet[37] have reported similar phenomenal behaviour maxima for different starchy materials and mixtures of flours and soya concentrates.

† A paper that is based on the same data, was presented at the Golden Jubilee Meeting of the American Society of Rheology in Boston 1979, and is to be published elsewhere.

The consistency developments of the raw material used, and of ground, water suspended extrudates of samples 6, 12 and 1 representing the mildest, intermediate and most severe 'treatments' in extrusion, respectively are given in Figs 2–5. Scanning electron micrographs (SEM) are displayed with their respective pasting curves. Figure 2 shows the consistency curves of the raw material. The solid curve shows the pasting behaviour without enzyme

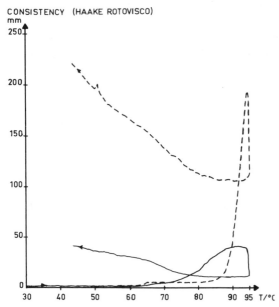

FIG. 2. Relative consistency in a pasting study as a function of temperature. The solid curve indicates the raw material pasting curve when amylolytic enzymes are not denatured. The dotted curve shows the pasting behaviour when amylolytic enzymes have been deactivated by using $0 \cdot 1 \%$ $HgCl_2$ and is the proper reference curve for extruded materials.

denaturation. The dotted curve shows the pasting behaviour when $0 \cdot 1 \%$ mercury chloride ($HgCl_2$) has been added in order to eliminate the effect of amylolytic enzymes on the raw material pasting behaviour. The latter curve is the proper reference for extruded products as no amylolytic enzyme activity survives in the conditions used in this experimental series.

When wheat flour, or other starchy raw materials, undergo heat induced denaturation–plastification during extrusion cooking, the initial cold water consistency of ground, water suspended products starts to rise, and the hot

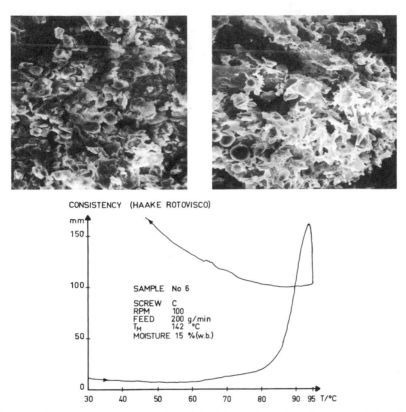

FIG. 3. Consistency curves and SEM pictures (magnification 300 ×) for the mildest treated sample, No 6 in Table 1. The SEM picture on the left shows the dry, ground extrudate. The SEM picture on the right shows the water suspended, freeze dried extrudate.

area peak consistency and 50 °C consistency upon cooling are reduced when compared with the raw material pasting curve with enzyme deactivation. The cold water suspended, freeze dried SEM pictures show that some starch particles have survived the mildest treatment, but that some extracellular (amylose) aggregate networks have already formed. The WAI value for the mildest treatment also shows an increase from the raw material WAI (Table 1).

When the 'hardness of treatment' increases, the initial cold water consistency of products increases further, and the hot area peak and 50 °C consistency upon cooling are further reduced until the maxima concurring

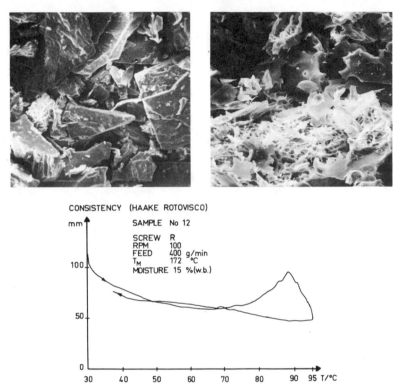

FIG. 4.   Consistency curves and SEM pictures (magnification $300 \times$) for the maximal initial consistency sample, No 12 in the series given in Table 1.

with the WAI maximum are reached. Cold water suspended, freeze dried extrudate SEM pictures show that this maximum area is connected with a sponge-like typical aggregate structure with few if any starch granules present. Even before the maximum area, and especially in areas further on the 'hardness of treatment', the ground, dry extrudates look like pieces of transparent plastic or glass in optical microscope and SEM observations.

When the 'hardness of treatment' further increases after the WAI maximum area has been reached, the initial cold water pasting consistency of ground extrudates starts to decrease while the hot area consistency and $50\,^\circ$C consistency upon cooling continue to decrease, until a pasting curve with low consistency throughout is obtained. The low consistency curve is accompanied by the disappearance of the spongy aggregate structures in the SEM pictures and high initial consistencies at the maximum. The

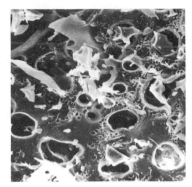

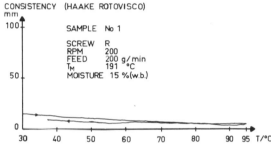

FIG. 5.   Consistency curves and SEM pictures (magnification 300 ×) for the most
severe treated sample, No 1 in the experimental series given in Table 1.

consistencies at the initial cold water state, hot area peak and 50 °C upon
cooling during pasting are given in Fig. 6.

The SEM pictures that are prepared by suspending the material in water
and freeze drying afterwards do not show the real structure of the material.
They are biased because of ice-crystal formation and subsequent changes
during drying.[22] However, the SEM observations in the above series, and in
other published and unpublished data, reflect changes in the aggregate
structures which are so large that the bias is small in its magnitude
compared with the changes due to progress of the treatment. So the
aggregate level changes can be considered to hold the key to understanding
the progress of denaturation and subsequent plastification of starches and
other biopolymers.

The microstructure of ground, dry extrudates is like that of plastic. If
similar biopolymer products are dried to low moisture contents close to 0 %
water, they display plastic-like behaviour.[34] For example, they have a

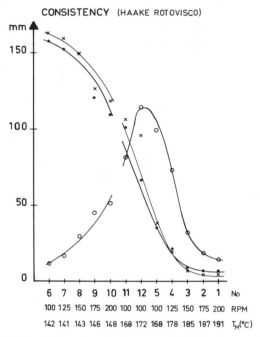

FIG. 6.   Initial 30 °C, ○, hot area peak, ×, and 50 °C, ●, consistencies for the experimental series in Table 1. Samples 6–10 are extruded with a screw combination having compression screws only. Samples 11, 12 and 5–1 have been extruded with a counter element in the screw composition.

softening or melting temperature zone. Thus the author suggests that the word 'melt' is a proper one to be used in connection with the plastifying mass in the extruder. This terminology suggests that the biopolymer–water mass is first formed into a dough, which upon denaturation and plastification is transformed into a melt during its passage through the extruder. At this point we can conclude that in our phenomena model the denaturation of biopolymers is a necessary prerequisite for plastification and texturisation of starch, proteins and cellulose alike.

The fact that at the aggregate level structural changes are responsible for the changes in functional properties needs further elaboration. Since the WAI experiments were performed at 30 °C, the water can be adsorbed by the molecular biopolymer structures and/or physically entrapped by the three-dimensional aggregate networks. What then causes the increased water solubility, decreased water absorption, and decrease in cold water consistency connected with the hard thermal processing in the extruder, or long cooking of starch under atmospheric conditions while shear is applied? In the author's laboratory only insignificant or nonexistent hydrolysis of starch to glucose or oligosaccharides has been found in enzyme denatured extrudates when applying twin-screw extrusion on cereal materials. There are indications that some depolymerisation of starch to the magnitude of 100–500 d.p. might occur, but this does not seem to explain the observed facts. We can again try to form a hypothesis.

2.4. A Hypothesis of the Reasons for Aggregate Structure Changes in Biopolymers upon Extrusion
When biopolymers are denatured, in that their quaternary structure is changed, the hydrogen bonds holding together the original aggregates are ruptured, and the macromolecules are released to form new hydrogen bonds. There is a possibility that chemical reactions proper also take place and their course also has an affect on the aggregate network formation. In starchy materials, for example, glycoproteins may be a part of dough formation and subsequent aggregate formation.[17] In proteins the disappearance and reappearance of S–S covalent bonds may also be a factor to be considered.

At mass temperatures around 170 °C something seems to happen in low water extrusion processing of starch and cellulose, and possibly of proteins as well. We can speculate that at this temperature changes in the tertiary structures of biopolymers start to take place predominantly, which leads to a loss of aggregate forming ability, and thus to changes in functional properties. Physically modified starches tolerate extrusion cooking better than native starches or flours. One should, however, remember that the physical conditions in extrusion are those of a 'reactor'. Thus the modification agents can immediately, after their release from the original cross-bonding, rebond with starch which may well lead to lowered denaturation of tertiary structures by protecting the helices associated with these structures.

2.5. Summary of Basic Changes in Extrusion Cooking
The dough formation–denaturation–plastification phenomena have been elaborated above, since the author thinks them to be a necessary basis for the understanding of extrusion cooking. 'Reaction analogous' in this connection refers to the fact that true chemical reactions such as hydrolysis are not the predominant mechanism. Instead changes on the aggregate level hold the key. However, the above hypothesis gives a perspective into the

understanding of the effects of heat, shear and ingredients such as mono-glycerides, fat and sugar on biopolymer denaturation and plastification.

If we consider the biopolymer, in this case starch, and water to be two 'reactants', we can say that ingredients do affect the 'reaction' of aggregate structure changes by

(a) influencing the state of water, or its 'reactivity' as sugar does,
(b) complexing immediately the aggregate forming molecules (amy-lose) being released by denaturation as monoglycerides or fatty acids of proper dimensions do, or
(c) affecting both the biopolymer and water as salt does.

Thus the 'reaction analogy' gives us a way to evaluate the effects of the recipe and ingredients, and their effect on the functional properties of products.

Chemical reactions proper, such as vitamin destruction, Maillard reactions between amino acids and carbohydrates, and to some extent dextrinisation or higher level depolymerisation of starch, also take place under the conditions encountered in extrusion cooking. Fortunately, their reaction rates seem to be lower than those of the aggregate physical changes, and so extrusion cooking yields a good retention of nutrients when applied in its normal food and feed processing conditions. Aggregate level changes, i.e. denaturation and plastification of biopolymers, and chemical reactions under these conditions are in principle irreversible. This means that the physical properties of products can be obtained from successive, irreversible changes taking place in the different zones of the extruder barrel, in the die and upon expansion at the die exit.

3. APPLICATION OF ELEMENTARY SYSTEMS ANALYSIS ON EXTRUSION COOKING

3.1. General

Figure 7 shows the basic principle of food processing systems. The food processing system (FPS)[40] consists of raw materials and ingredients, unit operations and processes, and the necessary mechanical equipment needed to produce a food item for the consumer. It takes into account the composition of materials used and the effects of processing conditions on defined quality criteria as they are evaluated by the consumer.

There are three major subsystems involved in the FPS. Subsystem 1, the phenomena level subsystem, relates the consumer quality to the raw materials and ingredient selection through the impact of basic changes during processing and storage. Subsystem 2, the mechanical equipment subsystem, consists of the hardware for the production and handling of a food item. It also forms the physical environment in which the basic changes during processing take place. Subsystem 3, the information and

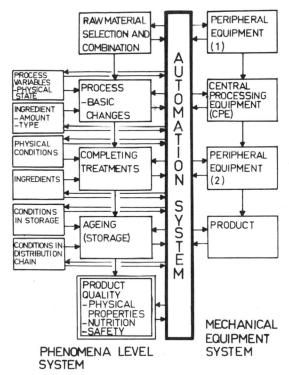

FIG. 7.   Schematic presentation of a food processing system (FPS).

automation subsystem, collects, processes and transmits information in a FPS, and executes the necessary control measures. It integrates the phenomena level subsystems and the mechanical equipment subsystems through instrumental measurements, information processing and process variable manipulation.

Inasmuch as the FPS essentially combines raw materials, the basic changes they undergo, unit operations and the properties of the product at the time of consumption, a special problem for the food industry is to predict and control consumer sensory quality factors, nutritional quality and safety already during processing, since the actual consumption may take place after an extended storage period.

Quality formation in processing is further redefined and simplified from the FPS concept (see Fig. 8).

The key to analytical and consumer quality formation is the interaction of product bases, or the raw material combination and processing conditions, on the basic phenomena. As the basic phenomena taking place in extrusion cooking are irreversible, the problem can be transformed to a general sequential optimisation problem and handled according to Bellman's principle of optimality for sequential systems[7] which states: 'An optimal policy has the property that, whatever the initial state and the

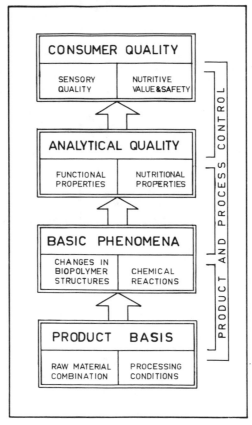

FIG. 8. Schematic presentation of food product quality formation in heat processing.

initial decisions are, the remaining decisions must constitute an optimal policy with regard to the state resulting from the first decision'. We shall now apply the sequential principle, which is the basis for the so-called dynamic programming, as a heuristic model for extrusion cooking.

3.2. Sequential Systems Analysis of a Cooking Extruder

Figure 9 shows a simplified principle drawing of a single-screw extruder. The mechanical central processing equipment (CPE in Fig. 7) in this FPS consists of a feeding device, a barrel that can be heated or cooled if the need arises, a screw rotating inside of the barrel, and a die through which the plastified mass re-enters the atmosphere. This chapter concentrates on

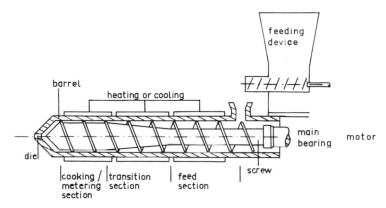

FIG. 9. Mechanical parts and functional sections of a single-screw extruder.

thermodynamically polytropic screw reactors which create the necessary physical environment, where the 'reaction analogous' basic changes in biopolymer aggregates and chemical reactions take place. The extruder is a high temperature, short residence time, relatively high shear, moderate to high pressure screw reactor.

Equation 1 (Section 2) describes the emergence of the overall functional and/or sensory properties in extrusion cooking as a function of raw materials, ingredients, moisture content and the effect of complete pressure, temperature, residence time and shear profiles in the CPE. This overall effect is, however, a combination of the effects in the different functional sections of the extruder. Figure 10 displays a heuristic, sequential model of the simplified extruder in Fig. 9.

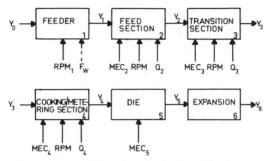

FIG. 10.   A heuristic, sequential model of the functional sectors for the single-screw extruder displayed in Fig. 9. $Y_0, \ldots, Y_6$ indicate the state of the quality vector at different stages of extrusion processing. RPM_1 indicates the rotation speed of the feeding device controlling the feed rate (F). RPM indicates the rotation of the extruder screw. MEC_2, MEC_3, MEC_4 indicate the mechanical construction of the screw and the barrel in the extruder feed, transition and metering sections respectively. Q_2, Q_3, Q_4 indicate the external heat added or removed in the extruder feed, transition and metering sections respectively. MEC_5 indicates the die geometry including its diameter, periphery and length. F_W indicates the water feed rate, if water is separately added to the feed section.

The quality property states in the sequential analysis of the extrusion cooker are described below.

$$Y_0 = f_0(RM, I, MC_0) \tag{2}$$

Y_0 is the raw materials and ingredients mixture at the moisture content it holds in the feeding device (MC_0). This implies knowledge of the type of raw materials and ingredients, and their ratios.

$$Y_1 = f_1(F + Y_0) = f_1(F + f_0(RM, I, MC_0)) \tag{3}$$

Y_1 is the amount of raw materials and ingredient mixture per unit time being fed into the extruder feed section. This amount is regulated in normal praxis by feeding device actions such as rpm of a feeder screw. In practical applications most extruders operate with partly starved screws, which means that the feed sections of extruders are not completely filled. The same applies for extruders equipped with flow restrictions such as steam locks also in other sections of the machine. Twin-screw extruders are always run in a partly starved condition. Thus the mass feed rate (F) is an independent process variable.

If water is added into the feed section separately, as often is the case, there is another independent variable, the water feed rate (F_W). In this case the total water content coming into the feed section is $MC = MC_0 + MC_1$ or the sum of moisture in the feed mixture and the moisture added separately into the feed section.

$$Y_2 = f_2(Y_1 + f_{PC2}(\tilde{P}_2, \tilde{T}_2, \tilde{R}_2, \tilde{S}_2)) \tag{4}$$

Y_2 is the state of the mass leaving the feed section. It is a function (f_2) of the entering mass, its moisture content, and feed rate, and including a function (f_{PC2}) of the changes that the pressure, temperature, residence time and shear profiles induce in this section. It is important to notice that the incoming mass and the extruder operation variables interact in the emergence of the $\tilde{P}, \tilde{T}, \tilde{R}, \tilde{S}$ profiles. Let us first consider the extruder operation variables in Fig. 10.

MEC_2 denotes the mechanical construction variables; the screw design inclusive flight depth, flight pitch, flight edge thickness, flight clearance and barrel surface construction (smooth or rifled). RPM is the screw rotation speed, and Q_2 is the amount of heat added or removed from this section, which in operational conditions is controlled by the barrel temperature. Feed sections are frequently cooled so that steam is not generated, since any steam in the feed section would result in uneven material flow and thus instability of the operation.

The mass that enters the extruder at feed rate F, is either a prepared dough or dry granular powder. If it is a granular mass it will at least be partly worked into a dough by the mixing and shearing actions of the screw and the temperature rise, given that sufficient moisture for dough formation is fed in. The entering mass has thus flow properties, which are changed by the dough formation. It also has a specific heat and heat transfer properties that are largely a function of the moisture content. Moisture also acts as a lubricant by reducing the viscous heat dissipation in the mass from the mechanical energy input through shearing action into the forming dough.

The temperature in the feed section can rise for either of two reasons. The first one is the viscous heat dissipation as the dough, if formed, is a highly viscous mass. Some ingredients such as bran material, and undissolved crystals such as sugar and salt increase the mass viscosity. They behave like glass beads in a solution and need material to cover them. The other mechanism leading to a temperature rise can be mechanical friction. Dry ingredients, if their particle size is so small that it allows them to enter the clearance inbetween the screw and the barrel, and if they are forced into this clearance, can create mechanical friction whereby more torque is needed and the mechanical energy is transformed into heat energy leading to a temperature rise. If sufficient moisture is present this friction-creating mass either gelatinises or water lubricates the surfaces thus solving the problem.

Without any mass entering the extruder through the feed section there is no formation of $\tilde{P}$, $\tilde{T}$, $\tilde{R}$, $\tilde{S}$ profiles. In summary the mass, its initial physical properties, and changes in the physical properties due to the mass–machine interaction create the final state $\mathbf{Y}_2$ of the mass leaving the feed section. At this stage the conditions or the state is seldom such that the initial denaturation temperature has been reached, and if it has been reached plastification has not yet substantially progressed. The normal maximal state of the mass in the feed section can be described to be a dough much like that in baking. The $\mathbf{Y}_2$ state is thus a mass with a set of physical properties leaving the feed section at a temperature that is higher than the initial incoming feed temperature.

$$\mathbf{Y}_3 = f_3(\mathbf{Y}_2 + f_{PC3}(\tilde{P}_3, \tilde{T}_3, \tilde{R}_3, \tilde{S}_3)) \tag{5}$$

$\mathbf{Y}_3$ is the state of the mass leaving the transition section. It is similarly formed from construction and operation variables, and the physical conditions interacting with the mass as was the state of the mass leaving the feed section ($\mathbf{Y}_2$).

The mechanical design is usually such that the mass inside the barrel compresses together by the physical action of the screw. This can be obtained by, for example, increasing the screw root diameter, which is equivalent to decreasing the flight depth, or by decreasing the flight pitch. The mass fills the screw fully in this section. Entrapped air, if present, is squeezed out from the mass by the rising pressure. The mass becomes a true continuum and the flow continuum mechanics start to function.

In a single-screw machine the mass is being pumped forward due to the action of so-called drag flow against the pressure from the decreasing space for the mass to employ, and the pressure introduced by the die. A single-screw extruder functions much like a centrifugal pump. The necessary condition for a drag flow to develop is that the mass has a higher adhesion to the barrel wall than it has to the screw element walls. If this is not the case, the mass will just turn around with the screw without being transported anywhere. Frictional surface in single-screw extruders is often increased by rifling the barrel surface. Rifling, however, also allows the mass to flow in a backward direction more freely. It also increases the heat transfer surface.

In many cooking extrusion applications the temperature in the transition section is allowed to rise above the necessary temperature for the initiation of the biopolymer denaturation and thus plastification sets in. This leads to drastic changes in mass physical properties connected with flow and heat transfer. In the initial stages after denaturation the biopolymers form new aggregate structures in the continuum. Thus, viscosity increases rapidly and viscous heat dissipation will also increase.

Two counteracting mechanisms in plastification set in. Mass viscosity will increase due to denaturation and availability of biopolymers for aggregate formation. At the same time the temperature rises due to viscous heat dissipation and often by heat flow from the extruder barrel wall. The newly formed aggregates are also subjected to shear. Viscosity of biopolymer masses is temperature dependent. It would decrease with increasing temperature if no more new aggregates were formed. It is also shear dependent; shear thinning and thixotropic with superimposed structural viscosity. In other words the biopolymer melts are rheologically strongly non-Newtonian. However, at early stages of denaturation and plastification the rise in viscosity is enormous due to the new aggregate formation in the mass continuum.

The viscosity of the biopolymer melt is a strong function of water content; the higher the water content the lower the viscosity. Also functional ingredients can interfere with the viscosity rise, for example, monoglycerides complex the otherwise aggregate-forming starch amylose, which results in a lower viscosity; sugar increases the starch gelatinisation temperature delaying the onset of denaturation by lowering the water 'reactivity', etc. So in practical technological applications complex situations can indeed arise.

Y_3 indicates the state of the mass leaving the transition section. In this heuristic model we take it to be a highly viscous dough, where the denaturation and plastification of biopolymers have started, and are proceeding with the aggregate formation causing a radical rise in the mass viscosity.

$$Y_4 = f_4(Y_3 + f_{PE4}(\tilde{P}_4, \tilde{T}_4, \tilde{R}_4, \tilde{S}_4)) \qquad (6)$$

Y_4 is the state of the mass leaving the metering section also called the cooking section. The stage limit can be drawn to the end of the screw flight, or through the void space between the screw end and the die entrance. If the construction of the extruder has a large intermediate space inbetween the screw end and the die entrance, it can either be handled as a separate stage, or connected to the die or to the metering section. Here we shall use the situation where the metering section is considered to end at the end of the screw flight.

Y_4 is again formed by the entering mass state (aggregate state, temperature, viscosity) and by the mechanical design variables of the screw and the barrel, and by the operation variables, RPM of the screw and heat flow indicated by the zone temperature. There is again the interaction of construction and operational variables, and the mass itself in the formation of $\tilde{P}, \tilde{T}, \tilde{R}, \tilde{S},$ profiles.

The basic changes, denaturation and plastification of biopolymers, which were discussed in connection with the transition section, continue in the metering section. The mass temperature in this section is usually the highest in the extruder. Aggregate formation and disruption continues. Viscous heat dissipation and often outside heating raise the mass temperatures into the range 100–240 °C in food and feed applications. Shearing can cause conformational changes in biopolymer tertiary structures in high temperature melts. We are now venturing towards the functional properties' maximum peak area, and, if heating and shearing are intensive enough, over it.

It was mentioned earlier that physical properties (viscosity, thermal properties) are a function of mass temperature. The same applies for reaction and aggregate plastification rates, and to the constants of the kinetic equations, as limiting factors change. When the mass viscosity reaches its maximum due to the new aggregate formation, there is a radical change in the trend; mass viscosity starts to decrease. We are over the peak. This maximum area, or the conditions connected with it, depends on the raw materials, the mass $\tilde{P}, \tilde{T}, \tilde{R}, \tilde{S}$ history, moisture content, etc. The following reduction of the melt viscosity is, in some cases, very rapid.

The surpassing of the maximum peak of melt viscosity does not always occur nor is it desired. If the extruder operates under conditions where the shear thinning, due to high shear rates between the flight edges and the barrel surface, and high barrel temperature lead to inproportionally large viscosity differences at the barrel surface and at the screw, slippage can occur which leads to instability of operation. If the mass temperature reaches values which are too high, burning can take place. This will happen at temperatures above 220–250 °C, depending on the product, the equipment used and on the operational parameters.

The mass that leaves the metering section is at least partly cooked, and normally in a plasticised state. Due to the fact that the pressure inside the extruder is high enough, water is in a liquid state despite the high temperatures. We can say that the mass acts like a polymer melt in its functional properties.

$$Y_5 = f_5(Y_4 + f_{PE5}(\tilde{P}, \tilde{T}, \tilde{R}, \tilde{S})) \qquad (7)$$

Y_5 is the state of the mass leaving the extruder die. Die functions include the shaping of the product and the creation of back pressure into the extruder thus affecting the residence time. In the case of fibrous materials such as textured vegetable proteins, the die functions include final orientation of fibre structure. Furthermore, the die induces additional shear into the plasticised mass.

There is always a void space between the screw end and the die, which is filled with plasticised melt. It can be a gap of a few millimetres or a larger reservoir with intermediate holed plates and channels of various constructions. This design detail has a large effect on the shear the mass experiences, and on the residence time in the extruder. In a twin-screw extruder, where no reverse flight elements exist in the screw design, the bulk of the shearing which the mass experiences takes place in the die and in the space between the screw end and the die entrance.

The die is a flow restriction, which creates a back pressure depending on the die design parameters of length, capillary diameter and periphery.[48] If special dies are used, as is the case in protein texturisation, the profile of the die cavity can have 90° channel direction changes. The back pressure which arises is a function of the mass viscosity, and changes during the time the mass resides in the die. Whatever is the pressure needed for the mass to flow through the die, it must be exceeded by the pressure, which is created by the pumping action of the screw or screws. The pressure created by the screws must also overcome the pressure loss in the void space before the die entrance.

The mass passing through the die is, in cooking extrusion applications, still pressurised, and plasticised to the extent of the sum of all the various sections of extruder actions, or the total of $\tilde{P}$, $\tilde{T}$, $\tilde{R}$, $\tilde{S}$ action. This mass re-enters the atmosphere through a restriction, the die, with speeds that can be rather remarkable under some conditions. The water in the mass is in a superheated state due to the elevated temperature used in many of the cooking extrusion applications. The water will thus violently boil. It flashes off in an instant, which leads to the expansion of the plasticised mass.

$$Y_6 = f_6(Y_5) \qquad (8)$$

Y_6 is the state of the mass after extrusion, which is a function of the state of the mass leaving the die exit. Given appropriate mass composition and appropriate conditions in the extruder, the state of the mass involves enough plasticised material for expansion. If the temperature of the mass is high enough for the water in the mass to boil, violent expansion takes place. Part of the expansion is die swell, which is due to the normal stresses that have been introduced into the mass by its viscoelastic behaviour in the die.

Proper explosive expansion needs, however, a plastifying raw material component and superheated water.

In food and feed extrusion cooking various natural and modified starches are needed as plastifying agents if expansion is desired. Proteins do not possess similar expansion ability, they tend to form laminar structures with less expansion. Proteins are also often cooked and extruded with higher water contents than starchy raw materials.

In order to be able to expand, starches must be denatured and the aggregate network worked to a proper state. If too much heat and shear is introduced into the plasticised starch mass, the state for the maximal expansion is surpassed. This is partly due to the state of the starch, inasmuch as the mass pressure is reduced, when plastification proceeds and the aggregate structures are 'worked over the peak' whereby not enough pressure differential is maintained across the die. Another factor is the lower mass viscosity in a starch mass that has been heated and sheared too much. In this case the stiffening of the plasticised mass happens too slowly in order to hold in the expanding steam bubbles and honeycomb structure. But given enough pressure differential across the die and a proper melt viscosity, starchy masses form a cellular structure when water flashes off at the die exit.

Various factors can interfere with the expansion operation. If there is enough nonplastifying material such as bran in the recipe, and the starch does not fully entrap the inert material in a manner where thick continuous melt layers exist, water is shot off from the structure including many vaporisation nuclei, and no expansion occurs. Monoglycerides do not allow aggregate networks to form, reducing the expansion if present in concentrations above 0.5%. If too much fat is added this also reduces expansion. Fatty acids of proper size complex amylose. Fat also acts as a lubricant in the melt thus reducing heating by viscous dissipation. If too much water is added it also acts as a lubricant, and requires too much thermal energy to vaporise enough of the water at the extruder die exit. This leads to a high moisture content viscoelastic mass that first expands, but due to too high a water content and high temperature after expansion (which leads to too high an elasticity) the initially expanded mass is redrawn to an unexpanded state.

The expanded mass that has been partly cooled upon the explosive expansion by the thermal energy needed to vaporise the water, is often in a viscoplastic state, and can be either cut or stretched and formed into continuous sheets as in the crisp bread production process. In snack food production the expanded mass is cut to pieces of desired lengths.

In some applications no expansion is desired at the die exit. This is achieved by cooking the starchy, or proteinous mass with high moisture contents, and cooling the mass below the boiling point of water while the mass still resides in the extruder, or by cooling in special die assemblies before the mass emerges into the atmosphere. The precooled, plastic mass can be dried and later expanded in an oil bath or microwave oven. Some of these masses have an interesting ground extrudate pasting behaviour. Even though the transparent, plastic like appearance shows that denaturation and plastification have taken place, the ground, water suspended extrudate shows a pasting behaviour much like that of nondenaturated starch. Figure 11 shows such an anomaly for barley starch.[58] The WAI value is also low.

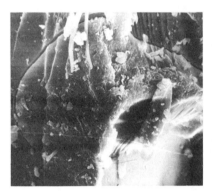 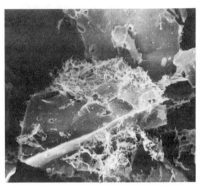

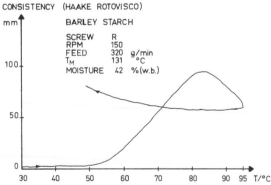

FIG. 11.   Consistency behaviour anomaly in pasting behaviour of barley starch, when moisture content during extrusion in a twin-screw extruder (displayed in Fig. 14) has been high. SEM picture on the left (magnification 1000 ×) is from dried, ground extrudate. SEM picture on the right (magnification 120 ×) is from the same extrudate, when it has been suspended in cold water and freeze dried.

This anomaly is due to the solubility and diffusional properties of the high water content mass. The plastified aggregate matrix is, in this case, extremely resistant to rehydration. When this mass is dried and later subjected to a rapid heating in an oil bath its expansion is due to the plastic like softening properties of the mass. This operation, where the remaining water expands the softened, less viscous mass, has been called alveolation by Menzi.[34]

We have now ventured through a single-screw extruder in a stage-wise manner. A similar systematic, heuristic model can be used to 'think any machine construction through'. There is not enough knowledge of the physical properties of denatured and plasticised masses, material–machine interactions, residence times in various zones of the extruder nor of the mass temperature and pressure profiles to afford exact mathematical modelling for each stage. Mathematical models of extrusion cooking will shortly be discussed below. Before that, however, let us inspect the differences of single- and twin-screw extrusion cookers.

4.   COMPARISON OF SINGLE- AND TWIN-SCREW EXTRUDERS

Single- and twin-screw extruders differ from each other most basically in their mass transport mechanisms. Single-screw extruders are like centrifugal pumps transporting the mass along the barrel by a mechanism of drag flow, which is dependent on the mass adhering more to the barrel wall than to the screw. Twin-screw extruders, on their part, act like positive displacement piston pumps. Janssen[24,25] and Martelli[32] discuss the action, benefits and problems associated with twin-screw plastics extruders. Clark[12] and van Zuilichem et al.[59] refer to the differences of single- and twin-screw extruders. The discussion below, will concentrate mainly on twin-screw, intermeshing flight extruders.

Due to the mass transport by positive displacement, mass transport action, twin-screw extruders introduce less shear into the mass than single-screw extruders. This leads to a smaller viscous energy dissipation in the mass. In some applications, where excess heat that has been generated through viscous heat dissipation is removed by cooling from a single-screw extruder, some heating is required in a twin-screw extruder. This is demonstrated schematically in Fig. 12.

Twin-screw extruders can have two modes of screw rotation; the screws can rotate in the same direction, in which case they are called co-rotating, or

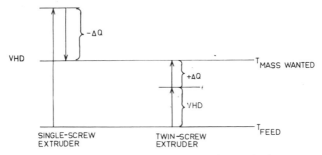

FIG. 12. Differences in temperature formation in a single- and twin-screw extruder. VHD is heat generated by viscous heat dissipation of mechanical energy, Q is heat removed or added by external heating or cooling.

they can rotate in opposite directions, in which case they are called counter-rotating. In a co-rotating machine the mass is pumped from one chamber of a screw to another chamber of the other screw in C shaped continuums. In a counter-rotating machine mass has to pass through a clearance between the flight top of one screw and the root of the flight in the other screw. Thus counter-rotation creates higher mass pressure formation in a shorter distance, but also introduces more shear into the mass and thus also more viscous heat dissipation. Figure 13 displays a principle drawing of mass pressure formation in a single-screw extruder, and in twin-screw co- and counter-rotating extruders.[1]

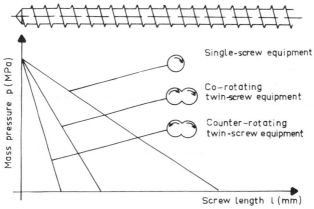

FIG. 13. Mass pressure formation in extruders for a single-, a twin-screw co-rotating and a twin-screw counter-rotating machine. (With permission, courtesy *Focus*, 1978.)

The counter-rotation tends to push the screws apart, when the mass is forced through the narrow gap between the flights of one screw and the screw roots of the other screw. Co-rotating machines are self-centering, and thus it is easy to design a self-cleaning screw system. The pros and cons for each of the twin-screw designs depends on the intended application.

Twin-screw extruders are at present relative newcomers on the food and feed extruder markets. It seems that machines with straight cylinders will be the standard twin-screw extruders in these industries.

Twin-screw extruder screws are designed and often manufactured in functional sections. The functions of each section can be analysed in a similar heuristic, sequential manner, which was used above for a single-screw extruder. Figure 14 displays a screw design, which is used in a twin-screw, intermeshing flight, co-rotating extruder. It also displays the assumed mass temperature and pressure profiles along the extruder.

If there is no counter pitched screw element in the screw design, the first flow restriction encountered by the mass is the flightless space between the screw and caps and the die entrance, and most importantly the die itself. The machine is filled just in front of the discharge end. In a case where a counter flight screw element is used in the screw design, this results in an effectively filled flow restriction, where the mass has to be pumped through the counter element by preceding compression screw action. Much more intensive shear is introduced, and the residence time in the extruder is greatly increased. This action by the screw can be designed to take place in the hot end of the extruder, as shown in Fig. 14, or it can be installed in the cooled section to induce effective mixing. Also other types of elements such as noncentred discs for grinding, or for protein dope molecular orientation can be used in the functional screw design.

The functional section design in twin-screw extruders, and the positive displacement material transport mode of these machines, offer great flexibility in the operations design. On the other hand this emphasises the need for systems analysis, and makes exact mathematical extruder modelling difficult.

A good example of modular functional design is given in Fig. 15. This design is used in a twin-screw, intermeshing flight, counter-rotating plastics extruder. In the opinion of the author this design is one of the most elegant constructions available. It has a conical barrel terminal section and increasing root diameter in the screw terminal sections. This allows well planned mass pressure, speed and temperature profiles at the final section of the extruder. The residence times for shear and temperature sensitive materials such as biopolymers can be made short, and the most wear-plagued parts, the end cone and screws, are short and thus cheaper to

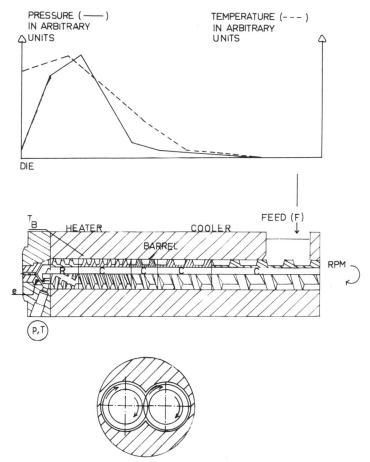

FIG. 14. Cross sections and assumed temperature and pressure profiles of a twin-screw, intermeshing flight, co-rotating extruder (Creusot-Loire BC-45). T_B, barrel temperature; RPM, screw rotation speed; C, compression screw element; R, counter pitched screw element; e, the gap between screw end and die insert; P and T, mass pressure and temperature probe location before the die.

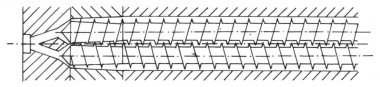

FIG. 15. Principle drawing of an e.s.d.e. twin-screw, intermeshing flight, counter-rotating plastics extruder. (With permission, courtesy *Focus* 1978.)

replace. Food extrusion could benefit from similar, innovative sectional designs.

Twin-screw extruders also have a great flexibility as reactors. This is a result of their mass transport mode and the positive displacement pumping action. They have been used in applications which involve decrystallisation of sugar for low water content boiled sweets, and similar designs have been used as crystallisers at low temperatures. Now applications of twin-screw extruders also include acid hydrolysis of cellulose which has been denatured in earlier sections of the one and the same machine,[2] and direct enzymatic hydrolysis of starch in the extruder.[29] In both cases the biopolymers are denaturated and hydrolysed in a single pass through the reactor. The author thinks that this is the beginning of screw reactor applications where few limits have been surveyed as yet.

A twin-screw extruder can also be applied in pelletising of problematic materials which have a high fat content. Here the fat melts enough to be extruded and solidifies upon the exit from the machine. Agglomerator applications are also being worked with.

Twin-screw extruders seem to be more versatile in their material handling ability and application ranges than single-screw extruders. One should, however, remember that they are relative newcomers in the food and feed processing industries. Many new research centres with new ideas and no concepts of 'the impossible' have entered the field with twin-screw machines. Their new applications will, in many cases, be transferred to single-screw machines with innovative designs. Which type of machine is the better for any given job depends on the product, peripheral equipment systems available, technology back-up and economics in each case.

5. MATHEMATICAL AND SYSTEMS MODELLING OF EXTRUDERS

We have so far only dealt with conceptual and heuristic models for the basic phenomena and extruder designs. They are beneficial in a science which is in its infancy, as extruder reactor research is. However, we need mathematical models to fill the 'function of' notations used in the previous sections. Before this problem is discussed, let us take a look at a hierarchical control system in Fig. 16.

Dynamic production planning is a matter of operations research and business economics. Steady-state optimisation of processes and direct

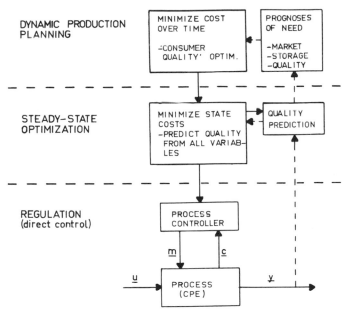

FIG. 16.   Schematic presentation of the principle of hierarchical control system.

regulation through controls are the hierarchy levels of engineers and technologists.

5.1. Basis of Modelling

Whenever we build a model, several basic things have to be considered:

1. Which dependable variables or variable combinations are the target of modelling?

2. Which variables are used as the independent variables, or the variables we take into account, when considering how to explain the way in which the dependent variables emerge from the real world?

3. What kind of a model are we going to build? Will it be based on fundamental physical and chemical entities, or will it be an empirical black or grey box model? Branch[8] has discussed this question, which most definitely influences the independent variable selection.

4. We should remember that any mathematical or heuristic model is but a reflected description of the real world. The purpose of modelling, and its usefulness in predicting the real world states, is a measure of the success of modelling.

The products of food and feed industries are intended to be eaten. Thus the final aim of the modelling, in this case, is given in the FPS—the product quality inclusive of functional properties such as sensory entities, safety and nutrition. The regular analytical techniques concentrate on functional properties such as water solubility index, water absorption index, expansion, consistency behaviour during pasting, mechanical strength, etc., or on chemically analysable entities. Here we are discussing a multi-dimensional quality vector, an example of which is given by van Zuilichem et al.[59] A lot more work has to be done in defining meaningful quality vectors for extruded food and feed products.

Other dependent variables that need modelling are the variables that give the product and process state anywhere along the processing path. These can be the mass pressure and temperature, viscosity, etc. at some point in the extruder. They are necessary as control variables. They should, however, ultimately be tied to the final product quality vectors in order to be useful. To emphasise the point, for a fish feed the decisive ultimate variable is the cost of feed used, or what it costs to gain a unit weight of fish. The pressure at the extruder head is not necessarily an indication of this factor. It can, however, be a necessary indicator for a good product in a proper state at this point.

As no model can be more accurate in its predictive strength than the accuracy of the analytical technique used, and many extruded products are not easy to analyse, there is a need to develop good indicators of product properties. Furthermore, quick, practical analyses for quality control to be used in the plant are worth considering.

5.2. Steady-state Modelling

Processes are run in a steady state. Thus steady-state modelling is used to predict chemical reactions, functional properties after processing, and state indicators during processing. There seems to be two apparently different philosophies employed. One is the empirical black box approach; the other one is the basic physical approach. Both have their merits and short-comings. Let us discuss the black box approach first.

The black box approach is based usually on response surface modelling (RSM). The principle is illustrated in Fig. 17. In this case the quality indicators are modelled from five independent variables, two originating from the 'recipe' and three processing variables. The second-order polynomial models for quality indicators are obtained through least squares fitting of the analytical data collected from an experimental design. This approach is beneficial when little if anything is known of the real form

RECIPE   PROCESS   QUALITY VARIABLES

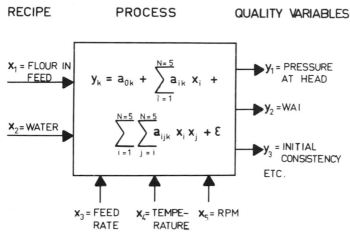

FIG. 17.   Principle of empirical black box response surface modelling (RSM) of quality or state indicators from recipe and processing variables. The equation in the box is a second-order polynomial in sum notation. ε = the residual after least squares fitting.

of the function in the box. One can inspect the effect of recipe and of process variables as they are really manipulated in the world on the end product quality indicators, and on the process state indicators at the extruder head where mass pressure and temperature can be measured.

These models strictly only apply to the ranges of independent variables used in the experimentation. A problem, which is also encountered when using polynomial models, is the difficulty that arises when physical and chemical interpretations are needed. On the other hand they can be applied to optimisation by linear programming[20] or simulation.[38] Applications of RSM in extrusion cooking are to be found in references 4, 5, 30, 31, 33, 44, 53 and 58.

The other approach is fundamental physical modelling. It should ultimately lead to the universal understanding and explanation of the $\tilde{P}, \tilde{T}, \tilde{R}, \tilde{S}$ profiles through the extruder and of their effects on the product quality indicators. The snag here is that the basic changes in biopolymers and the mass–machine interactions, and the changes in mass physical properties during extrusion are so complex that many simplifying assumptions have to be made. This leads to inaccuracy of the modelling. Besides, many of the 'constants' in the fundamental equations turn out to be functions of temperature or some other process variable.

The fundamental approach is, however, necessary, if extrusion cooking is really to be developed into a science, and if the knowledge so far gained in polymer extrusion is to be applied to food extrusion to its full benefit. This approach has been explored in references 10, 11, 16, 18, 19, 21, 24–26, 32, 39, 45, 46, 48 and 59. Pisipati and Fricke[45] have used the fundamental physical approach in sequential simulation of a single-screw extruder.

Tsao[54] and Sizer[49] among others have used first-order reaction kinetics and the Arrhenius equation in the modelling of nutrient retention in single-screw extrusion. This is an indication that reaction kinetics are entering into extrusion studies. More kinetic studies are needed for the extrusion processing conditions in order to take full benefit from the reactor approach to extrusion.

5.3. Dynamic Modelling

Dynamic models are needed for process control and stabilisation. Hereby the basis is a steady-state operation point and a given set of process state indicators at that point. Let us say that we have mass pressure and temperature as state indicators in an operations point indicated by screw rpm, barrel temperature, and feed rate for a given recipe. If any deviation is observed from the desired state indicators, controls take over and change a process variable so that the desired values of state indicators are reached. This presupposes dynamic models to be known. Olkku et al.[42] have performed a study on the dynamics and automation of a twin-screw extrusion cooker.

As biomaterials vary in their composition and behaviour in processing, and the dynamics of most extruders is unknown, much more work in this area is needed before the control applications fully meet the needs of expected automatic control techniques in extrusion cooking.

6. CONCLUSIONS AND COMMENTS

Extrusion cooking, which can also be called screw reactor processing, is one of the standard techniques employed in food and feed production today and even to a larger extent, in the future. The 1980s will see a stage in development where this technique is applied to an increasing number of industrial processes and will be experimented with in new, exotic fields. At the same time basic phenomena, their nonlinearity in physical space, and material–machine interactions are turned from mystery to manageable practice, even if there will still remain many problems to be solved.

Modelling of extruders is proceeding in its effort, volume and accuracy. Simulation techniques will be increasingly applied in both steady-state mathematical modelling and in dynamic modelling of extrusion processes. The necessary basis for this depends on the increasing understanding of the basic phenomena under screw reactor application conditions. These phenomena start with biopolymer denaturation at a given temperature, which is reached somewhere in the extruder barrel. This means that the total mass residence time in the whole equipment, which considers that the entire barrel is uniformly equal, is not a phenomenologically true approach. In twin-screw extruders especially the barrel is only sequentially filled by material.[41] So in order to master the extruder, one must consider the effect of various zones on the progress of the basic phenomena. Zones before the onset of denaturation have little if any effect on the plastification and subsequent texturisation. Those zones, which follow the area after the denaturation has started, affect the basic changes according to their pressure (P), temperature (T), residence time (R) and shearing effect (S) profiles. Here again the effective residence times in each zone have to be considered in order to follow the progress of denaturation and plastification phenomena. Thus residence time and mass accumulation profiles are needed for the whole passage through the extruder in order that they can be used in connection with the simulation models.

More detailed reporting of extrusion cooking data has to be emphasised. For example barrel temperatures reported are often but a gross indicator of the actual mass temperatures. Barrel temperatures which are more than 20 °C below the desired extrusion temperature have been found when extruding high moisture content 'non-frictional' raw materials. In some cases, where 'highly frictional' materials have been processed with low moisture contents, the observed mass temperatures have been as much as 60 °C higher than the set barrel temperatures. This throws some suspicion on the often reported 'extrusion temperature' unspecified data in the literature and on the validity of conclusions drawn on such data. For example one can ask the following question: 'Is the real effect of water in some reports really due to the changes of water concentration or is it due to resulting changes in extrusion mass temperatures?'

Plastics extrusion processing uses temperature controls within ± 1 °C of the desired set values. In some cases the screw design is altered when small changes are made in recipes. This is not always possible in food and feed processing, where the same machine has to be used often for many products, and recipe changes are a part of the daily praxis. We can, however, benefit from the accuracy of our more advanced colleagues in the plastics field by striving towards more accurate working schemes and modelling.

This chapter has not touched the nutritional aspects connected with extrusion cooking of foods. This important topic is worth an extensive state of the art review as nutritional value and safety are two of the main quality criteria for foods. The same applies to mathematical modelling, which should be thoroughly and extensively covered by experts in this field.

The basic and applied research effort in extrusion processing is presently increasing both in volume and in depth. It ranges from low cost extruders for field application to simulation and modelling of the most complicated twin-screw designs. Quality indicators and the basic phenomena are also being worked on. With good reason we can expect this to be beneficial in solving the problems of feeding the expanding world population and in problems connected with exploitation of renewing biopolymer resources.

DEDICATION

This chapter is dedicated to the memory of the late Arto Juhani Honkanen, who introduced the author to the art and science of extrusion.

REFERENCES

1. ANON., *Focus*, 1978 (3/4), 1.
2. ANON., *Kemisk Tidskrift*, 1980 (2), 55.
3. ANDERSON, R. A., CONWAY, H. F., PFEIFER, V. F. and GRIFFIN, JR., E. L., *Cereal Sci. Today*, 1969, **14**, 4.
4. ANTILA, J., 'Extrusion Cooking of Wheat Flour', MSc Thesis, Helsinki University of Technology, Espoo, Finland (in Finnish), 1979.
5. AQUILERA, J. M. and KOSIKOWSKI, F. V., *J. Food Sci.*, 1976, **41**, 647.
6. AQUILERA, J. M., KOSIKOWSKI, F. V. and HOOD, L. F., *J. Food Sci.*, 1976, **41**, 1209.
7. BEVERIDGE, G. S. G. and SCHECHTER, R. S. *Optimization: Theory and Practice*, McGraw-Hill Book Company, New York, 1970, p. 677.
8. BRANCH, J., *Chem. and Ind.*, 1975 (19), 832.
9. BRUIN, S., VAN ZUILICHEM, D. J. and STOLP, W., *J. Food Process Eng.*, 1978, **2**, 1.
10. CERVONE, N. W. and HARPER, J. M., *J. Food Process Eng.*, 1978, **2**, 83.
11. CLARK, J. P., *Food Technol.*, 1978, **32**(7), 73.
12. CLARK, J. P., *J. Texture Studies*, 1978, **9**, 109.
13. CONWAY, H. F., *Food Product Development*, 1971 (4), 27.
14. CONWAY, H. F., *Food Product Development*, 1971 (5), 14.
15. DONOVAN, J. W., *J. Sci. Fd Agric.*, 1977, **28**, 571.

16. Frazier, P. J., Crawshaw, A., Stirrup, J. E., Daniels, N. W. R. and Russel-Eggitt, P. W., In *Food Process Engineering. Vol. 1, Food Processing Systems* (Linko, P., Mälkki, Y., Olkku, J. and Larinkari, J., Eds.), Applied Science Publishers Ltd, London, 1980.
17. Graveland, A., Personal communication, 1979.
18. Harper, J. M., *Food Technol.*, 1978, **32**(7), 67.
19. Harper, J. M., *CRC Critical Reviews in Food Science and Nutrition*, 1979, **11**, 155.
20. Harper, J. M. and Wanningen, Jr., L. A., *Food Technol.*, 1970, **24**(5), 82.
21. Harper, J. M., Rodes, T. P. and Wanningen, Jr., L. A., *Chem. Eng. Progr. Symp. Series*, 1971, **67**(108), 40.
22. Hermansson, A.-M., Personal communication, 1980.
23. Horn, R. E. and Bronikowski, J. C., *Cereal Foods World*, 1979, **24**, 140.
24. Janssen, L. P. B. M. 'A Phenomenological Study on Twin Screw Extruders', Ph.D. Thesis, Krips Repro B.V., Meppel, The Netherlands, 1976.
25. Janssen, L. P. B. M. *Twin Screw Extrusion*, Elsevier Scientific Publishers Company, Amsterdam, The Netherlands, 1978.
26. Jao, Y. C., Chem, A. H., Lewandowski, D. and Irwin, W. E., *J. Food Process Eng.*, 1978, **2**, 83.
27. Lawton, B. T., Henderson, G. A. and Derlatka, E. J., *Can. J. Chem. Eng.* 1972, **50**, 168.
28. Lelievre, J., *Polymer*, **17**, 854.
29. Linko, Y.-Y., Vuorinen, H., Olkku, J. and Linko, P. In *Food Process Engineering. Vol. 2, Enzyme Engineering in Food Processing* (Linko, P. and Larinkari, J., Eds.), Applied Science Publishers Ltd, London, 1980.
30. Lorenz, K., Welsh, J., Normann, R., Beetener, G. and Frey, A., *J. Food Sci.*, 1975, **34**, 572.
31. Mannonen, L. 'Effect of Technical Parameters on Extrusion Cooking of Wheat Flour', MSc Thesis, Helsinki University of Technology, Espoo, Finland (in Finnish), 1979.
32. Martelli, F., *SPE J.*, 1971, **27**, 25.
33. Maurice, T. J. and Stanley, D. W., *Can. Inst. Food Sci. Technol. J.*, 1978, **11**, 1.
34. Menzi, R., In *Extrusion Cooking in Food. Cycle CPCIA Europe Séminaire E.6* (Mercier, C. and de la Gueriviere, J. F., Eds.), Paris, 1977.
35. Mercier, C., *Die Stärke*, 1977, **29**(2), 48.
36. Mercier, C., In *Food Process Engineering. Vol. 1, Food Processing Systems* (Linko, P., Mälkki, Y., Olkku, J. and Larinkari, J., Eds.), Applied Science Publishers Ltd, London, 1980.
37. Mercier, C. and Feillet, P., *Cereal Chem.*, 1975, **52**, 283.
38. Montgomery, D. C. and Bettencourt, Jr., L. A., *Simulation*, 1977, 113.
39. Morgan, R. G., Suter, D. A. and Sweat, V. C., *J. Food Process Eng.*, 1978, **2**, 65.
40. Olkku, J., *Kemia-Kemi*, 1978, **5**, 631.
41. Olkku, J., Antila, J., Heikkinen, J. and Linko, P., In *Food Process Engineering. Vol. 1, Food Processing Systems* (Linko, P., Mälkki, Y., Olkku, J. and Larinkari, J., Eds.), Applied Science Publishers Ltd, London, 1980.
42. Olkku, J., Hassinen, H., Antila, J. and Pohjanpalo, H., ibid.

43. OLKKU, J. and LINKO, P., In *Food Quality and Nutrition. Research Priorities for Thermal Processing* (Downey, W. K., Ed.), Applied Science Publishers Ltd, London, 1977.
44. OLKKU, J. and VAINIONPÄÄ, J., In *Food Process Engineering. Vol. 1, Food Processing Systems*, Linko, P., Mälkki, Y., Olkku, J. and Larinkari, J., Eds.), Applied Science Publishers Ltd, London, 1980.
45. PISIPATI, R. and FRICKE, A. L., ibid.
46. REMSEN, C. H. and CLARK, J. P., *J. Food Process Eng.*, 1978, **2**, 39.
47. ROSENBERG, K., 'HTST-Extrusion Cooking of Rye', MSc Thesis, Helsinki University of Technology, Espoo, Finland (in Finnish), 1979.
48. ROSSEN, J. L. and MILLER, R. C., *Food Technol.*, 1973, **27**, 46.
49. SIZER, C. E. 'Changes Occurring during the Extrusion of Potato Flakes', PhD Thesis, Colorado State University, Fort Collins, Colorado, 1976.
50. SMITH, O. B., In *New Protein Foods*, Vol. 2B (Altschul, A. M., Ed.), Academic Press, New York, 1976.
51. SMITH, O. B. and BEN-GERA, I., In *Food Process Engineering. Vol. 1 Food Processing Systems* (Linko, P., Mälkki, Y., Olkku, J. and Larinkari, J., Eds.), Applied Science Publishers Ltd, London, 1980.
52. SUZUKI, K., KUBOTA, K., OMICHI, M. and HOSAKA, H., *J. Food Sci.*, 1976, **41**, 1180.
53. TARANTO, M. V., MEINKE, W. W., CARTER, C. M. and MATTIL, K. F., *J. Food Sci.*, 1975, **40**, 1264.
54. TSAO, T. -F. 'Available Lysine Retention during Extrusion Processing', PhD Thesis, Colorado State University, Fort Collins, Colorado, 1976.
55. WILLIAMS, M. A., HORN, R. E. and RUGALA, R. P., *Food Eng. Int.*, 1977 (11), 57.
56. WILLIAMS, M. A., HORN, R. E. and RUGALA, R. P., *Food Eng. Int.*, 1977 (12), 23.
57. WINKLER, S., LUCKON, G. and DONIE, H., *Die Stärke*, 1971, **23**, 325.
58. VUORINEN, H., MSc Thesis, Helsinki University of Technology, Espoo, Finland (in preparation).
59. VAN ZUILICHEM, D. J., BRUIN, S., JANSSEN, L. P. B. M. and STOLP, W., In *Food Process Engineering. Vol. 1, Food Processing Systems* (Linko, P., Mälkki, Y., Olkku, J. and Larinkari, J., Eds.), Applied Science Publishers Ltd, London, 1980.

Chapter 7

THE EFFECT OF TEMPERATURE ON THE DETERIORATION OF STORED AGRICULTURAL PRODUCE

José Segurajauregui Alvarez†

and

Stuart Thorne
*Department of Food Science and Nutrition,
Queen Elizabeth College, University of London,
London, UK*

SUMMARY

Almost all agricultural produce, even that destined for processing or preservation by some other method, is stored in a fresh state for part of its post-harvest life. The reactions and processes that occur in such produce are many and complex and their overall effect manifests itself as quality change (usually quality loss) during storage. The rates and equilibria of these reactions are determined by many variables, particularly by temperature and atmospheric composition and humidity. Of these temperature has by far the greatest effect and fortunately it is the easiest to control.

It is not possible to consider the effect of storage temperature on the rate and equilibrium of each and every biochemical reaction. Such a model would be far too complex, some reactions are not completely understood and the mechanisms of some are unknown. It is necessary, therefore, to use gross simplifications and even the nebulous concept of 'quality' itself in order to

† Present address: Abedules 69, Jardines San Mateo, Naucalpan, Mexico.

produce working models of storage changes. In spite of this, many of the models proposed and evaluated in the literature can be used to predict, with adequate precision, the effects of temperature on storage changes in agricultural produce.

Produce is rarely stored at constant temperature. Methods are described which enable the effects of fluctuating temperatures on storage changes to be assessed.

1. INTRODUCTION

Most food products, no matter how they are processed, stored or preserved, deteriorate with time. For some products, notably those that are sterilised or dehydrated and properly packaged, the storage life is so long that it is often not of primary interest. But for other products, especially those that are frozen or preserved by cool storage in a fresh condition, storage life is a major consideration in the process. Of all the parameters that influence storage life, temperature is perhaps the most important and it is with the effects of temperature that we are concerned here.

Delays between the harvesting and utilisation of crops are inevitable and fresh fruits and vegetables are particularly prone to quality loss during the intervening period. Quality loss in produce destined for further processing can often be reduced by synchronisation of harvest, transport to the factory and processing. But the world's major food preservation process is fresh storage—90% of agricultural production is processed in no other way—and the distribution chain that ends with domestic consumption of 'fresh' produce is usually of unpredictable duration and subject to considerable changes in environmental conditions.

It has been estimated that between 25 and 80% of fresh fruit and vegetables are lost after harvest,[1] the most serious losses occurring in developing countries where they can least be tolerated. As Lorentzen[2] puts it; '... no matter how productivity is achieved, it will always be necessary to take good care of the harvest and bring it to the consumers with a minimum of quality loss.'

Individual products differ greatly in the length of time for which they can be maintained in an acceptable condition, in their responses to various treatments and to adjustments in environmental conditions. In general, the rate of deterioration of vegetables is related to the botanical function of the tissue. The shortest storage lives, and the highest metabolic rates, are found in tissues that were growing rapidly at the time of harvest, such as mushrooms, asparagus and other shoots. Next come leaves and other aerial parts of mature plants (cabbage, spinach, celery), followed by storage roots (carrots, turnips) and finally by specialised storage organs such as tubers (potatoes) and bulbs (onions). Table 1 shows typical storage lives and metabolic rates for a selection of vegetables related to botanical function.

TABLE 1
BOTANICAL FUNCTION RELATED TO STORAGE LIFE AND RESPIRATION
RATE FOR SOME PRODUCE

Product	Relative respiration rate	Botanical function	Typical storage life (weeks at 2°C)
Asparagus	40 ⎫	Actively	
Mushrooms	21 ⎬	growing	0·2–0·5
Artichokes	17 ⎭	shoots	
Spinach	13 ⎫	Aerial	
Lettuce	11 ⎬	parts of	1–2
Cabbage	6 ⎭	plants	
Carrots	5 ⎫	Storage	
Turnips	4 ⎬	roots	5–20
Beetroot	3 ⎭		
Potatoes	2 ⎫	Specialised	
Garlic	2 ⎬	storage	25–50
Onions	1 ⎭	organs	

Physiological deterioration in vegetables occurs primarily because of the stress induced by the removal of the tissue from the parent plant. The tissue is unable to replenish either the water or the metabolic substrates that it loses and may be unable to dispose of metabolic products. The moribund condition that results, mainly from excessive water loss, often predisposes the tissue to invasion by facultatively parasitic micro-organisms[3] and this condition is often exacerbated by mechanical damage during handling of the produce.

Deterioration in fruits is rather different in that it is closely related to the natural ripening and senescence of the fruit. For it is the natural function of the fleshy parts of fruits to deteriorate after the fruit leaves the tree and to release the seeds.

Considerations that apply to the effects of temperature on fresh vegetables can often be applied to frozen produce, although the temperatures involved are typically 25°C lower. Frozen foods in general present a simpler system than fresh fruits and vegetables. They are often blanched before freezing so that enzyme-catalysed reactions are of secondary importance, if they occur at all, and the main cause of deterioration may be comparatively simple reactions such as oxidation.

Important features of the environment which influence the longevity of fresh produce and which are amenable to control are temperature, humidity and the composition of the atmosphere surrounding the produce. Of these, temperature has the greatest effect on storage life and is, conveniently, the most readily controlled. Humidity is not readily controlled, but can have a considerable effect on deterioration; not only does it regulate the rate of water loss, but it has also been implicated in physiological storage disorders such as scald in apples.[4,5] Regulation of respiration by the control of atmospheric composition has been practised commercially for half a century since the pioneering work of Kidd and West.[6] Today, controlled atmosphere storage for apples and pears in the United States amounts to almost 400 000 tonnes capacity and in Britain to 250 000 tonnes.[7] Increasing awareness of the need to conserve produce has

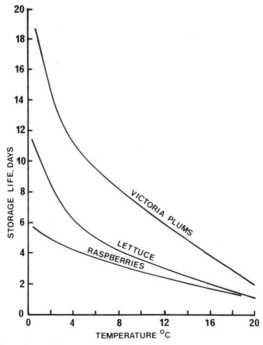

FIG. 1.   The effect of temperature on the storage life of some produce.

led to interest in the controlled atmosphere storage of other produce in recent years.[8,9]

But conservation of most fresh produce is effected solely by cooling it to the most suitable temperature as soon as possible after harvest, because low temperatures depress both the physiological activity of vegetable tissues themselves and of any micro-organisms capable of causing spoilage. The storage life is thus extended. Figure 1 shows the effect of temperature on the storage life of some produce and illustrates how considerable this can be.[10]

2. QUALITY

The criterion by which the acceptability of a food is judged by the consumer is that of 'quality'. Although an individual consumer will have no qualms about describing the quality of a food, quality is a complex function of many attributes of a food. For a fruit or vegetable, it will, typically, involve flavour and aroma, texture and firmness, size, shape and appearance. Many of these attributes are extremely difficult to measure quantitatively and, even if measurement is possible, the part that each plays in the overall concept of quality will vary from consumer to consumer. It is not, therefore, possible to develop a universal measurement of quality for any product. Not only does the relative importance of individual attributes vary, but many of them have limiting values that render the product unacceptable whatever the values of the other attributes. For example, a tomato could be unacceptable because it was too soft, even if its flavour and aroma, size and shape were perfect. Because of the difficulty of specifying quality, the concept of a 'quality indicator' is often used. This is the application of one or more attributes of a product to indicate storage changes. Ideally, a quality indicator should be a major quality attribute, but, even more important, it must be readily measurable with sufficient precision and reproducibility. With tomatoes, for example, firmness, surface colour and sugar/acid ratio have been used as quality indicators.[11] Other quality indicators proposed have included changes in vitamins[12] and internal colour of fruits.[13] The use of quality indicators is not satisfactory in that it does not give a real assessment of quality, but it is necessary in experimental work to utilise some objective and reproducible measurement rather than the nebulous and subjective concept of overall quality. The inherent variability of quality makes hypotheses based on it unreproducible. And the scientific acceptability of a hypothesis requires that it be reproducible at all places and at all times. So there is a strong case for the use of quality indicators although they are less realistic in practical terms. Nevertheless, direct measurement of quality—usually by sensory analysis—can give valuable practical data providing its limitations are realised. And much published information purporting to be based on quality measurement turns out, on close scrutiny, to be based on a quality indicator.

2.1. Prediction of Quality Change

Because temperature is the major factor influencing quality, it is very desirable to be able to predict the quality change of produce from known time–temperature data as a function of storage time. If the temperature is constant this can be done by experience, but transport and storage between the region of production and that of consumption usually involves great changes in temperature. Typical temperature histories of fresh and frozen produce are shown in Fig. 2, though it must be emphasised that these are most variable and often subject to unpredictable delays, especially fresh produce which has an indeterminate sojourn at the wholesale market and retail outlet.

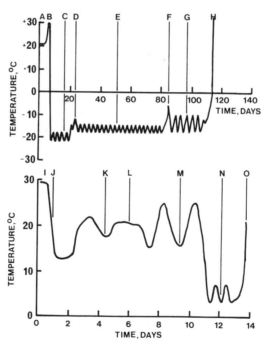

FIG. 2.   Typical temperature histories of fresh and frozen produce.

Different temperature histories induce different degrees of quality change, depending on the nature of the time–temperature pattern. Therefore it is necessary to develop a simple procedure to give an estimation of the accumulated quality change in a product at any time in the distribution system, no matter how variable or irregular the time–temperature pattern might be. This procedure should provide information on which to act for the protection of quality at any stage in the commercial life of the food. It should also provide information on the effects of individual operations in the distribution chain on quality; for example, on the effect of a nonrefrigerated journey or on the effect of leaving a pallet exposed to the sun for a short time.

Every analytical calculation needs a mathematical model. Various models have been proposed for studying the effects of temperature fluctuations on the storage of fruits and vegetables, both fresh and frozen. The 'effective temperature' theory, computer aided simulation models and the 'time–temperature-tolerance' hypothesis (TTT) are the most significant. Each of these models has partially fulfilled the prediction requirements already discussed. In particular, the effects of frozen storage on the quality of foodstuffs have quite successfully been predicted by the time–temperature-tolerance hypothesis.[14]

3. GENERAL CONSIDERATIONS AFFECTING POST-HARVEST CHANGES

As is true of most chemical reactions, the rate of enzyme-controlled reactions usually increases with temperature within the temperature range over which the enzyme is stable and retains full activity. Enzyme-catalysed reactions have an optimum temperature. The peak in a plot of catalytic activity against temperature occurs because enzymes, being proteins, are denatured by heat and become inactive as the temperature is raised beyond a certain point, usually between 55 and 60 °C.[15] The apparent optimum temperature is thus the result of two opposing processes; (i) the usual increase in reaction rate following the Arrhenius equation and (ii) the increasing rate of thermal inactivation above a critical temperature. So the effect of temperature on any of the individual enzyme-catalysed reactions which, together, effect storage deterioration is as illustrated in Fig. 3.

Physiological processes such as respiration, which are enzymic and can be used as general metabolic indicators, are affected by temperature in a similar way. Temperature coefficients of respiration, Q_{10} (defined as the

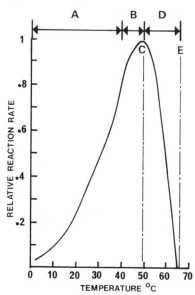

FIG. 3.   The effect of temperature on the rate of enzyme-catalysed reactions.

ratio of respiration rate at temperature $T°C$ to that at $T + 10°C$) increase less rapidly with increasing temperature than would be expected from the Arrhenius model. It has been suggested[16] that at low temperatures respiration rate is controlled by an enzyme reaction with a high Q_{10} value, whereas at higher temperatures it is limited by the rate of oxygen diffusion through the tissue, a process with a low Q_{10} value. Temperature coefficients of respiration also depend on the tissue age. But James[16] points out that, at higher temperatures, differences between temperature coefficients of similar tissues at different ages tend to disappear; evidence for oxygen diffusion limitation at these temperatures.

During post-harvest storage of vegetables or ripening and senescence of fruits, a great many metabolic changes occur simultaneously, for example, softening of the cell walls,[17] synthesis and destruction of pigments resulting in colour changes,[18] changes in carbohydrate composition and concentration resulting in changes in sweetness,[19] and destruction or polymerisation of phenolics resulting in reduction of astringency.[20] Some of these reactions are metabolically essential to the tissue. Some are secondary or by-products of the mainstream reactions. All have different temperature coefficients. If the fruits or vegetables remain attached to the parent plant or ripen in their natural environment, any possible imbalance of relationships of metabolic reactions resulting from fluctuations of temperature will be corrected by the tissue's hormonal system and by the interdependence of individual reaction rates on the product and substrate concentrations of other reactions. In other words, fruits will mature and senesce normally and vegetables will change only as required by the normal growth of the plant, free of physiological stress despite the metabolic changes taking place. But when the produce is separated from its natural environment, this ability to compensate for temperature change is severely impaired. Temperature changes can now result in the accumulation or depletion of metabolic substrates or products, resulting in considerable changes in those attributes that constitute quality.

When temperature fluctuation occurs, both equilibrium constants and the concentrations of the substrates involved are altered. For instance, the equilibrium constant for the hydrolysis of starch to glucose, a vital process, rises with temperature, i.e. high temperatures displace the equilibrium towards starch and low temperatures towards the sugars. Low temperatures also lead to the conversion of lipids into sugars and the accumulation of organic acids. It has been shown[16] that when potatoes are transferred from room temperature to temperatures below 8 °C, reducing and nonreducing sugars accumulate, which explains the respiration drift that they suffer when transferred again to room temperature. Upward changes of temperature are always accompanied by a temporary peak in carbon dioxide production when, presumably, these sugars are oxidised. Such effects are particularly important if the potatoes are to be used for frying for, in addition to flavour change, the sugar/starch ratio influences the degree of browning during subsequent cooking.

Work with apples and pears[21,22] has shown that a temperature transition from 1 to 20 °C is closely followed by an increase in carbon dioxide production, while production of ethylene—the hormone involved in ripening—rises to a peak after 5 or 6 h (before complete temperature equilibrium is achieved) and then falls. It has been suggested[21] that this increase in ethylene production is related either to the increased activity of an ethylene-producing system (which has developed at low temperatures due to the availability of substrates) or to tissue stress.

3.1. Extremes of Temperature

When produce, particularly fruit, is exposed to extremes of temperature, either high or low, physical and chemical disorders may occur; membrane permeabilities may be increased or reduced or some reactions may be accelerated or retarded to excess, resulting in permanent disruption of the normal ripening pattern. Due to under-concentration of essential substrates or to excessive accumulation of toxic products, physiological disorders obvious to sight or taste can ensue. A great variety of physiological injuries has been reported to occur in fruits subjected to chilling temperatures, for example internal browning in apples, skin spotting in citrus fruits, damage to vascular bundles in avocados, wooliness in peaches, and flesh breakdown in plums.[23]

Chilling disrupts the entire physiology of some sensitive fruits, particularly those of tropical or sub-tropical origin, and a number of mechanisms have been proposed to account for the effects. It seems likely that cell membranes undergo a physical phase transition at the critical chilling temperature from a normal, flexible liquid crystal to a solid gel structure. This change of state would be expected to bring about a concentration of the membrane components, causing the formation of holes and increased membrane permeability.[23] The phase transition may also increase the activation energy of membrane-bound enzymes, leading to interference with metabolic processes. Reduction in the supply of ATP coupled with the increased rigidity of membranes would account for the cessation of protoplasmic streaming which is one of the immediate symptoms of chilling injury.[24]

Exposure to chilling temperatures, unlike freezing which effects immediate damage, often needs to be prolonged before injury occurs. However, there are exceptions and bananas are damaged by even brief exposures to temperatures below 10 °C.[25] In general, the degree of chilling injury increases as temperature is lowered or exposure time is extended at any temperature below the critical one. In some cases, it has been observed that the effects of chilling are ameliorated if the tissue is subsequently transferred to a higher temperature.[26] Analysis has revealed that there is a higher proportion of saturated fatty acids in the lipids of chill-sensitive than of chill-resistant varieties.[23] It is also believed that membranes containing a high proportion of saturated fatty acids are less stable and more liable to undergo phase transition at chilling temperatures.

Just as abnormally low temperatures produce damaging effects, so do abnormally high temperatures. For instance, bananas fail to ripen above 30 °C;[27] they remain green and the pulp softens. Some varieties of plums show abnormal ripening at temperatures above 32 °C.[28] In avocados at 30 °C, as at 5 °C, the fruit does not ripen and the tissue darkens.[29] Lycopene synthesis in tomatoes is inhibited at temperatures above 30 °C[30,31] and ascorbic acid concentration decreases.[30] Furthermore, many fruits cease to produce ethylene between 35 and 40 °C[32] and as ethylene is a ripening hormone this has important results. In the tropics, temperatures are often higher than those required to produce these effects, but fruit on the plant usually recovers from exposure to very high temperatures.

4. THE TIME–TEMPERATURE-TOLERANCE HYPOTHESIS (TTT)

The rate at which any individual chemical reaction proceeds varies exponentially with temperature. Although the reactions and processes that effect quality changes in fruits and vegetables are many and varied, an exponential relationship to describe the overall effect of these reactions is still approximately valid, at least over moderate temperature ranges. Such a relationship can be used to relate storage life and temperature. Thorne and Meffert[33] fitted exponential equations to published temperature/storage life data for fresh and frozen fruits and vegetables and obtained very high correlation coefficients. Previously, several authors had reported similar relationships for a number of individual products.[34–38]

If storage life at any temperature is defined as the time required for various physiological changes to accumulate to such an extent that quality is just unacceptable, then the reciprocal of storage life is the rate at which these changes take place at that temperature. As an exponential curve can often be fitted to observed temperature/storage life data it should be possible to construct a curve of the rate of change (1/storage life) against time for any time–temperature history. The area under this curve up to any time would be a measure of the total change suffered by the produce up to this time. This is the basis of the method known as the time–temperature-tolerance hypothesis, first proposed by Van Arsdel.[14] It is analogous to the method first suggested by Bigelow et al.[39] and later modified by Ball[40] for determining the effectiveness of sterilising processes for canned foods; the universal method today. The 'lethal rate' of the Bigelow et al. and Ball method is analogous to the reciprocal of storage life in the TTT hypothesis.

The term 'storage life' requires acceptance of that dubious parameter 'quality' in its definition and therefore lacks nearly all the elements of a satisfactory quantitative measure. Nevertheless, it is possible to derive meaningful relationships between temperature and storage life, especially if a valid quality indicator can be used. Using such data, a graphical procedure for summing the effect of an irregular temperature history on quality or on a quality indicator can be used. The accumulative effect of a known temperature history upon quality can readily be estimated if reliable data on the rate of change of quality or of a quality indicator are available for several steady temperatures within the required range; provided that quality changes in the product are known to be additive and commutative and that there is no additional effect due to the temperature change itself.

It is unlikely that a quality indicator will be related linearly to temperature (or to time) and it must be linearised by a suitable mathematical function (exponential, logarithmic, etc.) so that the change of the property with time can be calculated from the slopes of these lines at different temperatures. In other words, the quality indicator, or a function of it, must depend only on temperature and never on the quality indicator itself; it must be explicitly dependent on temperature.

Working on the time–temperature-tolerance of frozen fruit and vegetables, Guadagni et al.[41] followed the rate of change of ascorbic acid in sliced strawberries and found that this decreased linearly with time at $-12\,°C$ and could, therefore, be used as an objective measure of deterioration during storage. Similarly, the logarithm of the chlorophyll content of green snap beans has been used as a quality indicator.[42] Segurajauregui Alvarez[11] has used surface colour and firmness to indicate quality in fresh tomatoes.

Approaches to a graphical solution of the time–temperature-tolerance hypothesis based on storage life data and on a quality indicator are illustrated in Figs 4 and 5 respectively. The former can be summarised by eqn (1) and the latter by eqn (2);

$$(1/L)_{\text{remaining}} = 1 - \int_0^t (1/L)\,.\mathrm{d}t \qquad (1)$$

$$Q = Q_0 + \int_0^t (\mathrm{d}Q/\mathrm{d}t)\,.\mathrm{d}t \qquad (2)$$

where Q is the value of the quality indicator at any time, Q_0 is its initial value and $1/L$ is the fraction of the storage life in the product. In the second of these approaches, a correlation has to be found between the quality indicator and the point at which the produce becomes unacceptable and there is a difficulty in identifying any quality attribute that might be a limiting factor for the produce. Complications in this procedure can arise if two or more deterioration processes are involved; and there usually are at least two. For instance, if the reactions resulting in texture change in a commodity have greater temperature coefficients than reactions causing colour change, then texture changes could limit storage life at higher

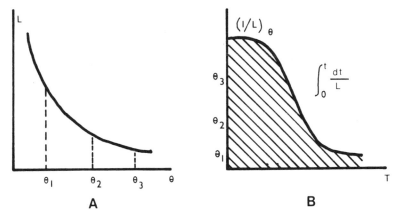

FIG. 4.   A graphical solution of the time–temperature-tolerance hypothesis (see eqn (1)).

storage temperatures whilst discoloration could be the limiting factor at low temperatures.

The final results of a TTT calculation can be expressed conveniently in either of two ways; (i) the length of time at a selected steady temperature (say 20 °C) that would produce the same change in the product or (ii) the effective steady temperature that would produce the same change in the produce in the same time.

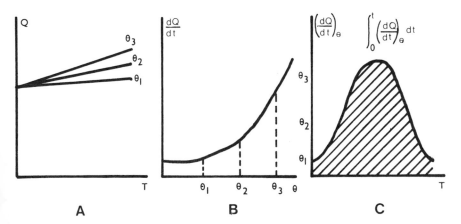

FIG. 5.   A graphical solution of the time–temperature-tolerance hypothesis (see eqn (2)).

FIG. 6.   Temperature history of tomatoes.

An example of the calculation will illustrate the time–temperature-tolerance hypothesis and its utility. Storage life (L) of green mature tomatoes can be related to temperature (T) by eqn (3):

$$L = 97 \cdot \exp(-0 \cdot 13T) \tag{3}$$

The effect on the quality of tomatoes of the temperature history illustrated in Fig. 6 is to be investigated. For small time increments along the curve of Fig. 6, the rate of quality change in the tomatoes ($1/L$) can be calculated from eqn (3). A curve of $1/L$ against time corresponding to Fig. 6 (shown in Fig. 7) can be constructed. The area under this latter curve is a measure of the total loss of storage life in the tomatoes. Integrating graphically between 0 and 10 days we get

$$\int_0^{10} (1/L) \cdot dt = 0 \cdot 87$$

i.e. 87 % of the tomatoes' storage life has been lost and 13 % remains. From eqn (3) it can be calculated that, for example, the time–temperature history for 10 days was equivalent to 6·7 days' storage at a constant 20 °C and that the tomatoes could be stored for a further 1 day at 20 °C without becoming unacceptable. (From eqn (3), $L = 7 \cdot 7$ days at 20 °C, $1/L = 0 \cdot 13 \, \text{day}^{-1}$.)

FIG. 7.   Ripening rate against time corresponding to Fig. 6.

As already mentioned, the TTT hypothesis rests on three important assumptions:[43] (i) additivity, (ii) commutativity and (iii) non-existence of additional effects due to temperature change itself. Additivity means that the total alteration in properties of a product, produced by a succession of experiences at various temperatures, is the simple sum of the separate amounts of change at each temperature and remains the same regardless of the number of breaks of continuity. Essentially, it means that changes of quality once produced in the food persist and are added to by all other changes that may be produced subsequently; that is, quality changes are irreversible. Commutativity means that the total alteration in properties is independent of the order of presentation of the various temperature experiences. Figure 8 illustrates these assumptions. Lines T_1 and T_2 represent quality changes with time at two different temperatures. Produce (I) is stored for 3 days at T_1 and then for 1 day at T_2; produce (II) is stored for 1 day at T_2 and then for 3 days at T_1. If the assumptions of additivity and commutativity are both satisfied, both lots will end up with the same quality after 4 days. Produce (I) will travel along the line T_1 for 3 days, cross to line T_2 without further quality change and proceed along this line for a further day. Produce (II) will proceed along line T_2 for 1 day, cross to T_1 and continue along this line for 3 days. If additivity and commutativity are valid assumptions, both produce (I) and (II) will end up with the same residual quality, as they do in Fig. 8. The non-existence of additional effects due to temperature change itself implies, in particular, that temperature gradients within the produce created when the ambient temperature changes have no effect.

These three assumptions have been tested in frozen foods for several commodities, for example strawberries,[41] sliced peaches,[44] raspberries,[45] orange juice concentrate,[46] peas and green snap beans,[47] spinach,[48] cauliflower,[49] turkey,[50,51] and chicken.[52,53] Charm et al.[54] have proposed a graphical method for estimating the shelf life of fish at any given set of storage temperatures based on these assumptions.

Because of the greater biochemical and metabolic complexity of fresh produce, application of the temperature–time-tolerance hypothesis is not easy and is often subject to limitations for individual products. Chilling injury is usually outside the scope of the hypothesis as it is neither additive nor commutative. But the method can often be applied within specified limits. Baxter et al.[55] have developed exponential equations to predict ripening of pears during and after cool storage by selecting firmness as a quality indicator. Segurajauregui Alvarez[11] has successfully used firmness and surface colour as quality indicators for ripening tomatoes but found

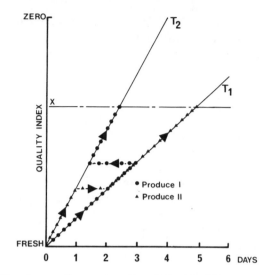

FIG. 8. Quality changes for produce I and II with different patterns of storage temperature.

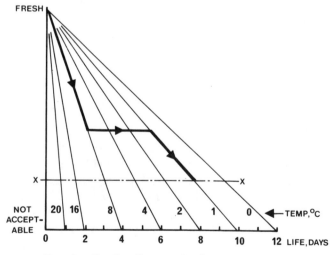

FIG. 9. Quality diagram for lettuce (see text).

that the time–temperature-tolerance hypothesis was only applicable if the tomatoes were not exposed to temperatures that caused chilling injury (i.e. below 12 °C). The Sprenger Instituut[56] presents temperature–quality data for fresh produce in such a way that it can be used as an approximate indicator of the effect of storage at more than one temperature. Superimposed on their quality diagram for lettuce (Fig. 9) is the effect of storing lettuce for 2 days at 8 °C followed by 2 days at 1 °C. The remaining quality (X) can be calculated from the diagram as being equal to $2\frac{1}{2}$ days storage at 0 °C, $1\frac{3}{4}$ days at 2 °C or $1\frac{1}{4}$ days at 4 °C.

Additivity and commutativity are implicit in the use of these diagrams and, although these have not formally been demonstrated for many products, such diagrams have been used successfully commercially. They are, of course, incapable of being used for predicting the effects of frequent or cyclic changes in temperature.

In general, it would appear that the assumptions of additivity, commutativity and the absence of temperature effects are valid for fresh fruits and vegetables, even during the ripening of climacteric fruits, unless drastic effects such as chilling injury occur.[11,55]

5. OTHER HYPOTHESES

Although the time–temperature-tolerance hypothesis, in its various guises, is the most widely applied theory to predict loss of quality in produce, a number of alternative theories has been proposed. The most important of these are the 'effective temperature' theory and various computer-aided simulation models.

5.1. The Effective Temperature Theory

The effective temperature theory proposes that if temperature varies with time according to some regular periodic wave then the extent of a particular reaction or change after a period of time will be the same as it would have been if the material had been held at some constant temperature for the same time. This theory was first proposed by Hicks,[57] who developed an equation for the estimation of the effect of sinusoidal diurnal temperature fluctuation on reaction rates in foods stored at ambient temperatures. It was further developed by Schwimmer et al.[58] who analysed different models of regular temperature fluctuation, for example saw-tooth, square and sine waves. Equations relating effective temperature to mean temperature and the wave amplitude temperature coefficient were developed for each model based on a zero-order reaction rate and on the following assumptions; (i) that the entire system fluctuated homogeneously so that analysis of one cycle would give data for the entire process, and (ii) that the wave amplitude temperature coefficient was constant throughout the temperature range of fluctuation and throughout the entire time.

The effective temperature theory has been widely tested for a number of frozen products in which deterioration is controlled by specific reactions, for example rancidity in turkey,[50,51] acid loss in strawberries,[41] oxidative browning in peaches,[44] migration of acidity from berries to syrup in raspberries,[45] rancidity in chicken,[52,53] retention of chlorophyll in snap beans,[42] retention of chlorophyll in peas,[47] and ascorbic acid retention and colour in cauliflower.[49] Repeated oscillations of as much as $\pm 5.6\,^{\circ}C$ ($\pm 10\,^{\circ}F$) within the range of -23 to $-1\,^{\circ}C$ (-10 to $+30\,^{\circ}F$) and cycle periods of 24–72 h were considered. In all cases studied, deteriorative effects on observed quality were, within the limits of experimental error, the same as they would have been after storage for the same length of time at the predicted steady temperature. Deteriorative reactions were independent of cycle frequency.

The limitations of the theory proposed by Schwimmer et al. were discussed by Singh and Wang,[59] and these limitations are even more important when considering the possible application of the model to the cool storage of fruits and vegetables. The effective temperature theory assumes a single kind of deteriorative reaction with a constant temperature coefficient. If two or more reactions with different temperature coefficients proceed simultaneously then the overall effect will not be predictable by the theory. Furthermore, it has been shown[16] that temperature coefficients depend on the temperature interval considered, their value decreasing as the temperature interval increases. In the case of fresh fruits and vegetables, temperature coefficients might also depend on the state of maturity and change during storage. Schwimmer et al. suggested that when Q_{10} varies with temperature, it should be possible to calculate the effective temperature by stepwise integration over the pertinent temperature ranges. They did not select the Arrhenius model for interpreting the relationship between temperature and reaction rate in frozen foods but considered an exponential model instead. If the logarithm of reaction rate constant is plotted against temperature for both the exponential and Arrhenius model, both follow approximately the same profile if the temperature range considered is fairly small.[60]

Another major assumption of the effective temperature theory is that the reaction rate is constant at constant temperature; this suggests a zero-order deteriorative reaction. The validity of this assumption is supported by Labuza[61] who points out that, although many foods deteriorate by first-order reactions, the difference between zero- and first-order is small except for very high degrees of deterioration. A limitation of the theory itself is that it is applicable only to cyclic temperature fluctuations. Cyclic fluctuations may occur in cool storage, but they are usually more complex than the simple functions which may be explained by the effective temperature theory. And the theory is unable to explain the non-cyclic changes that occur during transport and transfer to and from different stores, trucks, etc.

Although the validity of the effective temperature theory has been confirmed for a variety of frozen foods, its application to fresh produce has yet to be investigated.

5.2. Computer-aided Simulation Models

Computer-aided simulation of food quality for frozen and cool storage has been examined by various workers. The main objective of these studies has been to incorporate information on laboratory experiments on (i) the properties of foodstuffs, (ii) the kinetics of food deterioration reactions and (iii) the properties of packaging protection into a computer model of the entire system.

The extent of browning of frozen and dried cabbage stored in packages made of materials with different water vapour permeabilities was studied by Mizrahi et al.[62] The model that they developed was used in accelerated tests of browning in dehydrated cabbage.[63] Quality simulation models have also been proposed for products where two or more simultaneous deteriorative changes are important.[64] Quast et al.[65] investigated the storage life of potato chips (crisps), used as a model of a food that deteriorates by two simultaneous mechanisms; oxidation by atmospheric oxygen and textural changes due to water gain or loss. Kinetic information on ascorbic acid oxidation has been used in a computer simulation to predict quality changes in infant-feeding formulae during storage.[66] Factors such as light, dissolved oxygen concentration and package characteristics were included in the model to simulate quality.

Normally in frozen storage, a decrease in temperature is accompanied by an increase in stability of the product. However, some products show an opposite trend below a certain critical temperature. For example, bacon and liver accumulated more free fatty acids when stored at $-30\,°C$ than at $-12\,°C$. Singh[67] developed a computer-aided simulation model that incorporates the effects of storage temperature variations on the stability of frozen food products showing this reversed stability.

An important pre-requisite in computer-aided simulation of stored foods is the determination of kinetic information about deteriorative reactions. This kinetic data can then be incorporated into a program for a mathematical model that accounts for any variation in storage conditions. When several quality attributes are considered, the simulation should allow prediction of any one factor that limits quality; the consumer might, for example, reject a product with excessive colour deterioration even though all other quality attributes might still be at acceptable levels.

Analogue computer models for predicting physical and chemical properties of biological materials have also been investigated. Jabbari et al.[68] developed a model for apple fruit which predicts the change with time of starch, sucrose, glucose, fructose, malic acid, pyruvic acid, protopectin, soluble pectin, respiration rate and the total carbon dioxide output in storage under various atmospheric compositions and temperatures.

The utility and precision of computer models is limited by two major factors; (i) the model has to be constructed for an individual cultivar and (ii) a very considerable quantity of kinetic information is required to simulate an entire biological system.

6. CONCLUSIONS

Food systems are extremely complex and those involving fresh produce in particular involve many thousands of individual reactions, all with different temperature coefficients of rate and equilibrium. Furthermore, the rate of each reaction depends on the concentrations of its substrates and products, which in turn depend on the rate and equilibrium of other reactions. The only completely satisfactory way to predict the behaviour of such a system to different temperatures would be to consider the kinetic behaviour of every reaction involved. Not only would this approach involve a prodigious quantity of information, but detailed mechanisms of many reactions and even the reactions themselves are unknown. So practical information must be derived from simplified models based on such information that it is within human ability to obtain. Usually this means that, instead of considering individual reactions, it is necessary to consider the results of many reactions as if they were a single reaction; to consider colour changes or changes in texture for example. Such simplifications mean that there is no universal theory to explain the effects of temperature on deterioration and that there is a considerable empirical element in all theories. Nevertheless, such theories do provide useable information provided that their limitations are appreciated. These limitations usually involve a limited temperature range and restriction of a method to one specified humidity and atmospheric composition. Even when the limitations are observed, unpredicted failures of previously workable methods do occur occasionally because of lack of understanding of the kinetic detail of processes. But generally, if based on adequate experimental information, prediction is sufficiently precise for practical utility.

The most universally useful method for predicting the effect of various temperature histories on produce is the time–temperature-tolerance hypothesis. It is the simplest method, but also the method most removed from fundamental changes in the food. Great care must be taken to ensure that time–temperature-tolerance data is applied only to the cultivar for which it was derived, that conditions of atmospheric composition, humidity, etc. are not changed and that the data are not extrapolated to temperatures beyond the original experimental range. Given these limitations, the time–temperature-tolerance hypothesis, in its various guises, is a workable theory.

Effective temperature theory is one step nearer to considering fundamental changes, but it is limited to conditions of regularly cycling temperature. Computer simulations can come nearest to a fundamental approach, provided that adequate and sufficiently accurate information can be provided; if such information is inaccurate it can easily induce greater inaccuracies than the gross assumptions of the time–temperature-tolerance hypothesis.

It is doubtful that even a theory based on fundamental data would be completely foolproof, for it could never account for all variations that occur from batch to batch even with the same cultivar of any one fruit or vegetable. Behaviour depends on, among other considerations, time of planting and harvest, soil and weather conditions. Post-harvest storage behaviour can also vary significantly from one end of a field to the other!

Prediction of the effect of temperature on stored produce is imperfect. Doubtless methods will improve and doubtless they will never be perfect. But provided the limitations of methods are appreciated, useful practical infomation can be obtained.

REFERENCES

1. McGLASSON, W. B., SCOTT, K. U. and MENDOZA, J., *Int. J. Refrigeration*, 1979, **2**(6), 199–206.
2. LORENTZEN, G., *Int. J. Refrigeration*, 1978, **1**(1), 13–26.
3. THORNE, S. N., *J. Sci. Fd Agric.*, 1975, **26**, 933–40.

4. KIDD, F. and WEST, C., *Rep. Fd Invest. Bd for 1932*, 1933, 58–62.
5. KIDD, F. and WEST, C., *Rep. Fd Invest. Bd for 1933*, 1934, 199–204.
6. KIDD, F. and WEST, C., Food Investigation Special Report No. 30, HMSO, London, 1927.
7. RYALL, A. L. and LIPTON, W. T., *Handling, Transportation and Storage of Fruits and Vegetables*, Vol. 2, AVI, Westport, 1974, pp. 341–3.
8. PARSONS, C. S., ANDERSON, R. E. and PENNEY, R. W., *J. Proc. Am. hort. Soc.*, 1970, **95**, 791–4.
9. TOMKINS, R. G. *MGA Bull.* 1966, **201**, 477–8.
10. Sprenger Instituut, *Produktgegevens Groente en Fruit*, Sprenger Instituut, Wageningen, 1974–1980.
11. SEGURAJAUREGUI ALVAREZ, J., Ph.D. thesis, University of London, 1980.
12. KRAMER, A., *Fd Technol.*, 1974, **28**, 50.
13. BIRTH, G. S. and NORRIS, K. H., *The Difference Meter for Measuring Interior Quality of Foods and Pigments in Biological Tissue*, USDA Technical Bulletin No. 1341, Washington DC, 1965.
14. VAN ARSDEL, W. B., *Fd Technol.*, 1957, **11**, 28–33.
15. LEHNIGER, A. L., *Biochemistry*, 2nd edn Worth Publishers, New York, 1975.
16. JAMES, W. O., *Plant Respiration*, Oxford University Press, Oxford, 1953.
17. PILNIK, W. and VORAGEN, A. G. J., Pectic substances and other uronides. in *The Biochemistry of Fruits and their Products*, (Hulme, A. C., Ed.), Academic Press, London, 1970, pp. 53–87.
18. SCHANDERL, S. M. and LYNN, D. Y. C., *J. Fd Sci.*, 1966, **31**, 141.
19. BIALE, J. B., *Advances in Food Research*, 1960, **10**, 293–354.
20. GOLDSTEIN, J. L. and SWAIN, T., *Phytochem.*, 1963, **2**, 371–83.
21. TORRES, M. A. and RHODES, M. K. C., *Agronomia lusit.*, 1973, **34**(4), 347–59.
22. LONGHEAD, E. G. and FRANKLIN, E. W., *Canadian J. Fd Sci.*, 1971, **51**(2), 170–2.
23. LYONS, J. M., *A. Rev. Pl. Physiol.*, 1973, **24**, 445–66.
24. RAISON, J. K., *Bioenergetics*, 1973, **4**, 285–309.
25. WILLS, R. B., *Pl. Physiol.*, 1975, **56**, 550–1.
26. SMITH, A. J. M., *J. hort. Sci.*, 1950, **25**, 132–44.
27. GANE, R., *New Phytol.*, 1936, **35**, 383–402.
28. UOTA, M., *Proc. Am. Soc. hort. Sci.*, 1955, **65**, 231–43.
29. BIALE, J. B. and YOUNG, R. E., *Endeavour*, 1962, **21**, 164–74.
30. TOMAS, M. L., *Bot. Gaz.*, 1963, **124**(3), 180–5.
31. OGURA, N., NAKAGAWA, H. and TAKEHONA, H., *J. agric. Chem. Soc. Japan*, 1975, **49**(4), 189–96.
32. WILKINSON, B. G., Physiological disorders of fruit after harvesting, In *The Biochemistry of Fruits and their Products*, (Hulme A. C. Ed.), Academic Press, London, 1970, pp. 537–54.
33. THORNE, S. N. and MEFFERT, H. F. TH., *J. Fd Quality*, 1979, **2**(2), 105–12.
34. KIDD, F. and WEST, C., *Proc. Roy. Soc.*, 1930, **106**, 93–109.
35. TINDALE, G. B., HUELIN, F. E. and TROUT, S. A., *J. Dept. Agric. Vict.* 1938, **36**, 18.
36. ULRICH, R., *Proceedings of the XVII International Congress*, Vol. 3, 1967, pp. 471–94.
37. MORRIS, L. L., *Ice Refrig.*, 1947, Nov., 41–2.
38. TOMKINS, R. G., *Rep. Fd Invest. Bd*, 1937, 184.

39. BIGELOW, W. D., BOHART, G. S., RICHARDSON, A. C. and BALL, C. O., *Heat Penetration in Processing Canned Foods*, National Canners' Association, Washington DC, 1920.
40. BALL, C. O., *Univ. Calif. Publs publ. Hlth*, 1928, 1, 15.
41. GUADAGNI, D. G., NIMMO, C. C. and JANSEN, E. F., *Fd Technol.*, 1957, 11, 389–97.
42. DIETRICH, W. C., NUTTING, M. D., OLSEN, R. L., LINDQUIST, F. E., BOGGS, N. M., BOHART, G. S., NEUMANN, H. J. and MORRIS, H. U., *Fd Technol.*, 1959, 13, 136–45.
43. VAN ARSDEL, W. B., *Quality and Stability of Frozen Foods, Time–Temperature-Tolerance and its Significance*, Wiley-Interscience, New York, 1969.
44. GUADAGNI, D. G., NIMMO, C. C. and JANSEN, E. F., *Fd Technol.*, 1957, 11, 33–42.
45. GUADAGNI, D. G., NIMMO, C. C. and JANSEN, E. F., *Fd Technol.*, 1957, 11, 633–7.
46. McCOLLOCH, R. J., RICE, R. G., BAUDURSKI, M. B. and GEUTILI, B., *Fd Technol.*, 1957, 11, 444–9.
47. BOGGS, M. M., DIETRICH, W. C., NUTTING, M. D., OLSEN, R. L., LINDQUIST, F. E., BOHART, G. S., NEUMANN, H. J. and MORRIS, H. U., *Fd Technol.*, 1960, 14, 181–5.
48. DIETRICH, W. C., BOGGS, M. M., NUTTING, M. D. and WEINSTEIN, N. E., *Fd Technol.*, 1960, 14, 522–7.
49. DIETRICH, W. C., NUTTING, M. D., BOGGS, M. M. and WEINSTEIN, N. E., *Fd Technol.*, 1962, 16, 123–8.
50. KLOSE, A. A., POOL, M. F. and LINEWEAVER, H., *Fd Technol.*, 1955, 9, 372–6.
51. HANSON, R. J., *Fd Technol.*, 1958, 12, 40–3.
52. HANSON, R. J., *Fd Technol.*, 1959, 13, 221–4.
53. KLOSE, N. S., *Fd Technol.*, 1959, 13, 474–84.
54. CHARM, S. E., LEARSON, R. U., RONSIVALLY, L. U. and SCHWARTZ, Z., *Fd Technol.*, 1972, 26, 65–8.
55. BAXTER, R. I., BEATTIE, B. B. and HALL, E. G., Tech. Pap. Div. math. Statist., CSIRO, Australia, No. 36, 1972.
56. Sprenger Instituut, *De Houdbaarheidsformule*, Report No. 1853, Sprenger Instituut Wageningen, 1973.
57. HICKS, E. W., *J. Coun. Scient. Ind. Res. Aust.*, 1944, 17, 111–14.
58. SCHWIMMER, S., INGRAHAM, L. L. and HUGHES, H. M., *Indust. Engng Chem.*, 1955, 47, 1149–51.
59. SINGH, R. P. and WANG, C. Y., *J. Fd Process Engng*, 1977, 97–127.
60. KWOLEK, W. F. and BOOKWALTER, G. N., *Fd Technol.*, 1971, 25, 1025–32.
61. LABUZA, T. P., *J. Fd Sci.*, 1979, 44, 1162–8.
62. MIZRAHI, S., LABUZA, T. P. and KAREL, M., *J. Fd Sci.* 1970, 35, 799–803.
63. MIZRAHI, S., LABUZA, T. P. and KAREL, M., *J. Fd Sci.*, 1970, 35, 804–7.
64. QUAST, D. G. and KAREL, M., *J. Fd Sci.*, 1972, 37, 679–83.
65. QUAST, D. G., KAREL, M. and RAND, W., *J. Fd Sci.*, 1972, 37, 673–8.
66. SINGH, R. P. and HELDMAN, D. R., *Trans. Am. Soc. agric. Engrs*, 1976, 19(1), 178–84.
67. SINGH, R. P., *Int J Refrigeration*, 1978, 1(2), 108–12.
68. JABBARI, A., MOSHENIN, N. N. and ADAMS, W. S., *Trans. Am. Soc. agric. Engrs.*, 1971, 14(2), 319–25.

Chapter 8

THERMAL STERILISATION OF FOODS

L. W. WILLENBORG

*Stork Amsterdam B.V., Amsterdam,
The Netherlands*

SUMMARY

In this chapter a range of aspects related to sterilisation will be discussed. First of all, we will take a look at micro-organisms and at the techniques we can employ to inhibit their growth or destroy them altogether. This is followed by some observations on the mathematical methods needed to compare processes with one another. Next we will discuss the term 'cooking', in connection with the sterilisation of food products, and the containers used for such products.

The two factors, product and container, determine the choice of sterilising system, to which we will devote a few explanatory notes before examining the Stork Hydromatic steriliser in some detail.

The chapter is concluded by a schematic overview and a look into the not too distant future, for which we must plan today.

1. THERMAL STERILISATION OF FOODS

Sterilisation by means of heat is one of the methods available to preserve food products and give them a longer shelf life. Other techniques that can be employed to preserve foods are irradiation, drying, deep freezing, concentration, salting and smoking.

All these processes are designed to prevent micro-organisms from utilising the medium presented by the food. They will either keep the organisms from growing or destroy them completely. It is important, however, that the preservation process should have the least possible impairing effect on the product.

The keeping quality of food products depends not merely on the presence or absence of micro-organisms, but also on the enzymes within the food product. Enzymes play the rôle of biochemical catalysts that cause food products to deteriorate and start decomposing. Thus preservation has the added object of stopping this degradation process by inactivating the enzymes present in the food.

The inactivation of both micro-organisms and enzymes can be brought about by heat, a fact that was discovered by Louis Pasteur around 1880. The pasteurisation process he invented and lent his name to enables a product's shelf life to be prolonged by a few days. Nowadays, it still finds application in some areas, such as the dairy industry.

To prolong a product's shelf life beyond a couple of days, all thermotolerant or thermoduric organisms in the product must be killed, and the enzymes inactivated.

Thermal food processing necessitates some knowledge of how micro-organisms enter and multiply in food and of the consequences. The organisms that concern us most are those which cause food spoilage and those which are harmful to man. The micro-organisms found in food are moulds, yeasts and bacteria. Of these, bacteria present the greatest hazard.

Moulds occur everywhere in nature. They are found in the soil, on vegetation and even in the air around us. They multiply through spore formation, which entails a kind of chain reaction, i.e. they develop spores, which grow into moulds, which then produce spores again, and so on.

Fully grown moulds cannot withstand heat treatment, but some types develop spores which are highly resistant to heat and which are known as thermoduric. They can survive exposure to 92 °C for more than 1 min.

Like the moulds, yeasts occur widely in nature. They are highly useful in certain processes, such as fermentation, and the production of beer, alcohol and glycerol. Yeasts can reproduce through sporulation, but they are also capable of multiplying by division or fission. They grow at temperatures ranging from 20 to 35 °C and, as a rule, cannot withstand heat treatment at the temperature needed for pasteurisation (approx. 90 °C).

Bacteria are the third group of micro-organisms we have to deal with. They consist of a single cell and are among the smallest life forms known to man. They occur in a wide variety of shapes, e.g. round (cocci), elongate (rods), twisted (spirilla), etc. and reproduce only by direct cell division or fission. The growth rate of bacteria is subject to a range of factors, which includes the species of bacterium involved. Some bacteria also produce spores, but in contrast with those of yeasts and moulds, such spores constitute a phase of dormancy in the normal life cycle. When conditions become unfavourable for their continued growth, these bacteria develop spores which are extremely resistant to heat, cold and chemical agents and are therefore a means of ensuring the organism's survival.

Since bacteria are living organisms, they need food for their subsistence and reproduction. This food consists of nitrogen combined with organic matter.

Another crucial element in the life of bacteria is oxygen. Aerobic bacteria can only grow in the presence of oxygen and anaerobic bacteria will only multiply in the absence of oxygen.

A further determinant of bacterial growth is the ambient temperature.

For the purpose of thermal sterilisation, bacteria can be classified as follows:

Psychrophiles, with optimum growth between 14 and 20 °C
Mesophilos, with optimum growth between 30 and 36 °C
Facultative thermophiles, with optimum growth between 35 and 46 °C
Thermophiles, with optimum growth between 50 and 60 °C

Bacterial life is shown schematically in Fig. 1.[1]

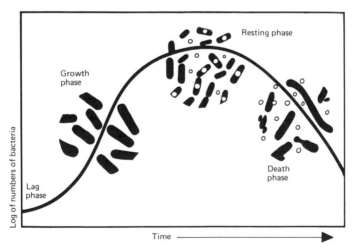

FIG. 1. Typical growth rate curve for an organism placed in a suitable growth medium. Accumulation of the growth by-products causes the organism to enter a stationary phase and subsequently the drying off phase, in which the vegetative cell disintegrates completely. If the organism is a spore-forming bacterium, only its free spores will remain intact. (Reproduced with permission.[1])

At high acidity levels (pH values below 4·6) bacterial growth is hardly possible and although pasteurisation will not destroy any bacterial spores, this process can be applied successfully in the preservation of such acid foods as fruit, sauerkraut, pickles, some seafood products, etc.

All these products have a comparatively low pH that prevents spore growth and, consequently, product spoilage.

Sterilisation has the purpose of eliminating all microbiological activity in a given product, whilst minimising impairment of the product's quality. The process is carried out at temperatures ranging from 110 °C to as high as 140 °C. It can be used on most meat products, vegetables, babyfood, petfood and liquid dairy products, and will give them a commercial shelf life up to one year.

2. MICROBIOLOGICAL ASPECTS

The destruction of microbiological spores can be illustrated by TDT (thermal death time) graphs. These are semilogarithmic graphs, in which the time is shown on a vertical log-scale and the temperature on a horizontal linear scale. By means of experiments, the decimal reduction time, D, which is the time required to eliminate 90% of a given spore type, can be established for several temperature constants and then plotted on a TDT graph.

Since the death rate of spores is represented logarithmically, the curves obtained will have the form of straight lines in the semilog TDT graph. Thus for any given micro-organism, the graph gives the relation between the times and temperatures at which 90% of the organism's spores will be killed.

2.1. TDT Graph for Two Different Types of Organisms (Fig. 2)

The lines result from heating tests conducted at various temperatures to determine the decimal reduction time D. A TDT curve provides another useful index, known as the Z value. The Z value is the number of degrees centigrade required for the TDT curve to traverse one logarithmic cycle. It is a measure of the change in death rate with a change in temperature. In Fig. 2, the Z values for organisms A and B are indicated, respectively, by Z_A and Z_B. A third quantity frequently used in thermobacteriology is the F value. It represents the amount of time needed to destroy a given number of micro-organisms at a given temperature.

As the thermal resistance of microbiological spores depends heavily on

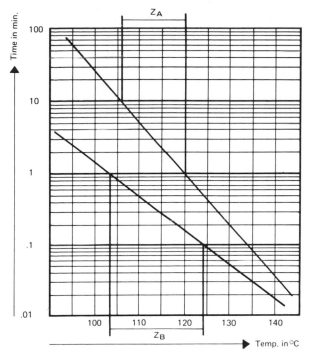

FIG. 2. TDT graph for two different types of organisms.

the type of product they are contained in, it has been necessary to introduce a standard unit of lethality, called the F_0 value. It is based on a decimal reduction time (D value) of 1 min at 121·1 °C and a Z value of 10 °C.

The F_0 value of a process is the value resulting from a constant temperature of 121·1 °C applied for a certain number of minutes, based on instant heating and cooling of the product involved. At any other constant temperature (T), the F_0 value is found by the following formula:

$$F_0 = t \times 10 \exp\left\{\frac{T - 121·1}{10}\right\}$$

in which t is the time in minutes and T is the temperature of the product during time t.

With the help of the F_0 value as a unit of lethality, we can compare processes conducted at different temperatures for different periods of time. Also, it enables processes involving a gradual temperature increase and decrease to be evaluated. In practice it is used as a standard by which to rate

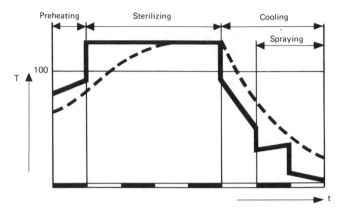

FIG. 3.   Time/temperature graph. The horizontal axis reflects the time and the vertical axis the temperature of both the sterilising medium (———) and the product (— — —). The F_0 value of the process can be calculated by integrating the temperature in excess of 100 °C through the formula given above.

the lethality of a process, and the minimum value adopted for most products is seven. The F_0 value can be established through graphs or by chemical analysis, but the general point of departure is always the time/temperature graph of the sterilising process shown in Fig. 3.

2.2.   Time/Temperature Graph (Fig. 3)
Nowadays there is electronic measuring equipment capable of making such graphs automatically. A probe placed at the slowest heating point in the container registers the temperature at specified intervals. The temperatures measured are stored in a memory and can be retrieved by means of a printer. The printer also provides a direct reading of the F_0 value, whilst the market even offers sophisticated equipment that registers the ambient temperature as well as the internal and external container pressure next to computing the F_0 value.

3.   THE SIGNIFICANCE OF STERILISED FOOD

Over the years, the social function of sterilised food has undergone a marked change. Initially, sterilisation was adopted for the purpose of rendering summer food surplusses fit for consumption during the winter months. One example of such sterilisation is the traditional process of preserving food in jars. It can be regarded as the forerunner of commercial food sterilisation. It consists of heating products to 95–98 °C in glass jars provided with a rubber sealing ring. When the product and the water vapour cool down, the resulting vacuum creates a hermetic seal. Food preservation in jars was (and is) a form of domestic industry designed to cover private needs.

The start of the 20th Century saw the introduction of commercial food preservation. The first canneries were built and processed such foods as peas, beans, carrots, and the like. This young industry was largely season-oriented and highly labour intensive. Sterilised food acquired a different rôle in that it had to supply the needs of an industrialising society with its rapidly expanding cities. The initial period of commercial food preservation ended by 1950. Since then the food processing industry has made great forward strides that resulted in today's modern food plants with outputs, for example, of up to 1800 containers per minute and running 6000 h per year on a round-the-clock basis. The product range has become much wider and now includes complete instant meals, specialities, diverse meat and fish products, soups, liquid and solid infant food, and pet food.

The variety of container types and materials has been broadened and today preserved food is available in cans, glass jars, aluminium pouches, plastic containers and other packages.

From our brief historical survey it might be inferred that the evolution of the food processing industry runs parallel with a growth in affluence and could even be used as a gauge by which to compare countries in terms of the prosperity they have achieved. Lending scientific support to this thesis is outside our present subject. The description given merely serves as a general framework in which we should place today's food processing industry and the aspects to be broached later on, when we will deal with the equipment used in sterilisation.

4. STERILISATION OR COOKING

Given the great number of products and container types, it is obvious that each product requires its own specific heat treatment.

As has been explained in Section 2, the adoption of elevated temperatures involves far more rapid destruction of microbiological spores and a much slower rate of chemical change in the product. This explains the prevailing tendency to increase processing temperatures and thereby reduce the processing time, so that product quality will be improved. However, a limiting factor in applying higher temperatures is the transfer of heat from the periphery to the slowest heating point in the container. Such transfer becomes even more complicated in products containing solid particles, because here the overall transfer rate depends on the partial transfer rates from the container wall to the viscous phase $\alpha 1$, from the viscous phase to the solid particle $\alpha 2$, and inside the solid particle $\alpha 3$. Thus the total heat transfer coefficient is determined by several factors, namely:

Viscosity of the product
Agitation of the container
Percentage of headspace in the container
Difference in density between liquid and solid particles
Ability of solid particles to move freely in the liquid phase

4.1. Heat Transfer Inside the Container (Fig. 4)

Rotation seems to be the best method of increasing heat penetration. However, the use of rotation involves several aspects that call for prior examination.

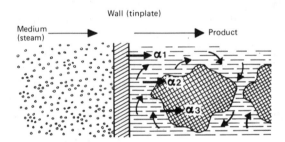

FIG. 4.   Heat transfer inside the container.

The effect of rotation is largely conditional on the amount of headspace available in the container, the viscosity of the product handled, the ratio between liquid and solid fractions in the product, and the size of the solid particles.

In a container the air bubble formed by the headspace acts as agitator. The larger the bubble, the higher the agitating effect at a given rotation speed.

The heat treatment applied to a product is determined by the slowest heating point in the container, because by working from this point we ensure complete exposure to the treatment.

For each individual product it is necessary to establish whether it effectively undergoes sterilisation, based on the requirement that the slowest heating point should reach a given F_0 value (see Section 2) or a given cooking level. Examples of products for which the lethality (F_0 value) is employed as a criterion are milk, green peas and the like, whilst the cooking level serves as a yardstick for meat products, soups, and beans in tomato sauce. Rotation is applied specifically in cooking processes, for the purpose of accelerating heat transfer. However, just as with other forms of food sterilisation, such rotary processes must invariably be based on an F_0 value of seven being reached at the slowest heating point inside the container, even when the effect of rotation is nil through the absence of headspace. Furthermore, the effect of rotation depends on the number of revolutions and the rotation radius.

Another factor to be considered is that rotation may inflict damage on discrete particles. Examples in point are beans in tomato sauce and whole peeled tomatoes.

Thus, while under well defined conditions, such as those found in laboratory tests, rotation may help improve product quality; it is not the ideal way to obtain a better product when applied industrially under conditions that often involve unknown variables. Also, rotation entails the risk of understerilisation resulting from the so-called 'solid pack' effect.

The drawbacks presented by rotation have stimulated the search for quality improvement through better product control and more insight into such facets as lethality and cooling effect.

High temperature/short time processes affect the balance between microbiological inactivation on the one hand, and enzymatic inactivation on the other. The Z value of microbiological spores is much lower than that of enzymes. This implies that HTST processes may cause enzymatic inactivation to lag behind, so that although products thus treated have a better quality just after the process, their shelf life will be much shorter.

5. PRODUCT QUALITY

For a long time the quality of heat sterilised food was judged by purely subjective standards. Since the early 1970s, however, a greater interest has been taken in finding qualifications more accurate than 'commercially sterile' and including such aspects as colour, texture and general appearance in the assessment. Most likely this interest was brought about by new types of sterilisable packages and increasing competition among food processors.

As indicated, while rotation may drastically increase heat penetration, the need for a given cooking level usually implies a certain amount of oversterilisation to ensure that every single container will receive an adequate heat treatment.

As shown in Fig. 5, treating a product in containers of a flat shape takes far less time, even considering safety margins, than processing it in cylindrical containers.

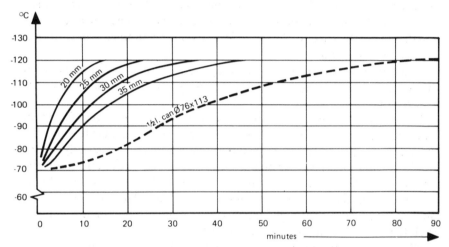

Fig. 5. Heat penetration curves for white beans in tomato sauce packed in 1/2-litre cans (76 × 113 mm) and in flat pouches (120 × 190 mm of varying thickness). Filling temperature = 70 °C; steam temperature = 123 °C.

As early as 1954, Guyer and Holmquist[1] have referred to the enzymatic activity in HTST-processed foods. In 1974, Yamaguchi and Kishimoto[2] indicated that the shelf life of HTST-processed products may be considerably shorter than that of products sterilised by a conventional technique and stored at the same ambient temperature. This fact was already known in the dairy industry, where UHT-processed milk is taken as having a shelf life of only 8–12 weeks, whereas conventionally sterilised milk may keep for three to six months under the same storage conditions. The difference in question is due to the fact that the inactivation of certain enzymes by a short-time process at ultrahigh temperatures is less effective than the inactivation brought about during the much longer processing time required at 118–121 °C.

Thus food sterilisation not only involves the elimination of micro-organisms (as reflected in the F_0 value) but also adequate inactivation of enzymes (notably peroxidase).

Some data on the behaviour of thiamine (vitamin B_1) and certain enzymes have been published by Reichert,[3] although the figures he provides represent relative rather than absolute values. Just as for the thermal destruction of micro-organisms, enzymatic inactivation and thiamine retention can be shown by straight lines in semilog graphs (see Fig. 6) and resolved through the Arrhenius equation:

$$F = t \cdot 10 \exp\left\{\frac{T - T_{ref}}{Z}\right\}$$

whose symbols are explained in Table 1.

The influence of thermal processing conditions relative to these factors is illustrated for a few cases, all based on an F_0 value of 10 min for the slowest heating point.

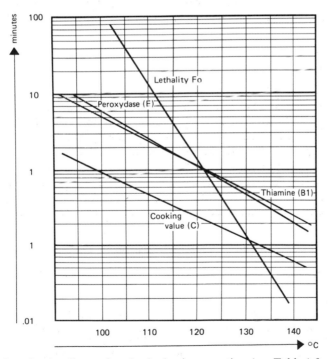

FIG. 6.   Inactivation lines using the Arrhenius equation (see Table 1 for values).

TABLE 1

F	T_{ref}	Z	Dimension
Processing value (F_0)	121·1	10	minutes
Cooking value (C)	100	33	minutes
Peroxidase destruction (E)	121·1	30	minutes
Thiamine (B_1) retention	121·1	26·1	%

The C, E and thiamine retention values have been integrated for the entire product, since the container periphery is exposed to more intensive thermal treatment than the inside.

Here again, the figures should be seen as relative rather than absolute.

When we apply this approach to commercial products in different container types, we can calculate the various values with the help of the time/temperature graph in Fig. 7. The figures speak for themselves, and although they merely represent a calculation model they make plain that to derive the maximum benefit from the potential for rapid heat penetration offered by flat containers, we must adopt other yardsticks besides the F_0 value.

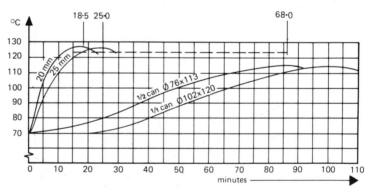

FIG. 7. Calculated results of different thermal processes in hydrostatic sterilising.

Package	T1	Preheating (min)	T2	t	F_0	C	E	B_1 retention (%)
1/1-litre can, 102 × 119	70	15	123	90	10	190	44	26
1/2-litre can, 76 × 113	70	15	123	71·0	10	168	37	30
Pouch, 20 mm	70	5	130	13·5	10	54	13	61
Pouch, 25 mm	70	6	130	20·0	10	68	16	52

In doing so, we should not lose sight of the fact that the quantities concerned cannot be expressed directly in terms of product quality, flavour, and similar attributes. They are only suitable for use in comparing thermal processes applied to identical products.

This concludes our theoretical exposition on the sterilisation of food. We have seen that micro-organisms show different behaviour patterns, that the inactivation of enzymes is as essential as the destruction of micro-organisms, and that both aspects can be translated into measurable quantities. One crucial factor which remains to be considered is the container.

6. CONTAINERS USED IN FOOD STERILISATION

The primary function of the container is to prevent food from being infected by the re-entry of micro-organisms after sterilisation has been completed. Next to this, the container has a number of secondary functions that may vary with container type. The oldest type of container used in industrial food processing is the can, invented by Peter Durand in 1810. The first cans were heavily tin-plated and sealed by soldering prior to sterilisation. They have no resemblance to today's cans, which consist of two pieces, are deep-drawn or have a single side-seam and come with varnished lids.

The container must be sealed hermetically to preclude any infection. Since sterilisation is performed at temperatures varying from 110 to 135 °C, a temperature gradient will occur between the filling stage and the sterilising stage, which entails a pressure build-up in the container. Needless to say the amount of pressure a container can withstand depends on its design.

The pressure developing in the container is the result of several factors, chief among them being water vapour pressure. Most food products to be sterilised contain a fairly large amount of water, and this water will cause a certain pressure in the container as the processing temperature increases. In fact there is a direct relationship between temperature and water vapour pressure. The pressure increase resulting from the water vapour pressure corresponds to the difference between the water vapour pressure at the filling temperature ($Pvap_{T1}$) and the water vapour pressure at the sterilising temperature ($Pvap_{T2}$). This pressure increase is depicted in Fig. 8.

Another factor responsible for pressure increase in the container is the air entrapped in the container's headspace, which will expand as the product is heated from filling temperature to sterilising temperature. Mathematically, the resulting pressure increase follows the equation $(PV)/T = C$, according

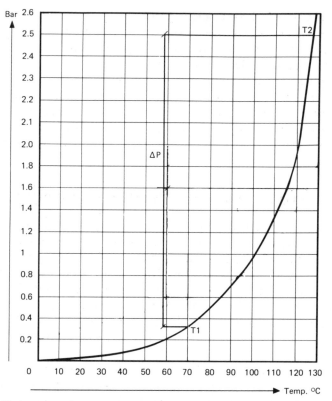

FIG. 8. Temperature/water vapour pressure relationship. The temperature is shown on the horizontal axis and the water vapour pressure on the vertical axis. Example: The pressure increase at a filling temperature of 70 °C and a sterilising temperature of 127 °C is $Pvap_{127} - Pvap_{70} = 1$ bar.

to the law of Boyle and Gay-Lussac, but in practice it is hard to establish with accuracy, because the container can itself absorb a certain increase in volume.

The third factor underlying the pressure increase arising in the container is the expansion the product undergoes as a result of the increase in temperature.

Aside from these three factors, which occur invariably regardless of the product, there may be an additional pressure increase caused by the release of gases through chemical reactions in the product. The solubility of gases in the product may change with variations in temperature and this will lead to a higher pressure in the container.

Thus the overall pressure that will develop in the container during the sterilising process can only be calculated on an approximate basis. This pressure is reflected in the formula:

$$Ptot = Pvap + Pair$$

The factor $Pvap$ is found by the subtraction $Pvap_{T2} - Pvap_{T1}$ where $Pvap_{T2}$ = water vapour pressure at the sterilising temperature, and $Pvap_{T1}$ = water vapour pressure at the filling temperature.
The factor $Pair$ can be determined by the formula:

$$Pair_{T2} = Pair_{T1} \times \frac{V_{T1}}{V_{T2}} \times \frac{T_2}{T_1}$$

where
$Pair_{T1}$ partial air pressure during the sealing operation, which is usually 1 bar, unless container sealing takes place under a vacuum.
V_{T1} headspace at filling temperature $T1$
V_{T2} headspace at sterilising temperature $T2$
$T2$ sterilising temperature in degrees Kelvin
$T1$ filling temperature in degrees Kelvin.

V_{T1}, $T2$ and $T1$ are measurable and can be found empirically. V_{T2} depends on the following variables: (a) the container's elastic deformation, (b) the expansion of the product, (c) the expansion of the container.

When the container is a glass bottle, as employed in the sterilisation of, say, milk or infusion fluids, the elastic deformation is negligible and the factor V_{T2} can be calculated. Owing to this absence of deformation, glass bottles can withstand very high pressures.

When the container is a can, V_{T2} practically defies calculation, because the deformation at the bottom and at the lid depends upon the external and internal pressures which produce a state of equilibrium.

It is this equilibrium which ultimately determines V_{T2}. Hence the load on the can's seam is limited to the factor Pw, which equals P external $- P$ internal.

In a flexible container, such as a pouch or a plastic bottle, a similar state of equilibrium Pw will occur. As the resistance inside such containers is virtually 0, we can say that P external $= Ptot$ internal.

Here again, it is possible to calculate V_{T2} and $Ptot$. The three variants referred to are shown in Fig. 9. When we plot the pressures on a graph, the differential pressure Pw becomes apparent for the various container types during the various stages of the process (see Fig. 10).

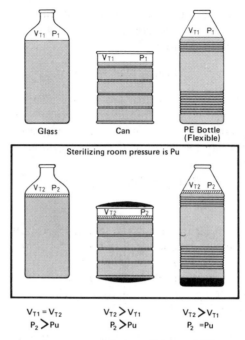

FIG. 9. The three container types under conditions of filling and sterilisation. In the figure, the hatched area represents the volume increase resulting from expansion, and the black area represents the volume increase caused by elastic deformation.

It is evident, therefore, that the container employed plays a vital role in the sterilisation of food products.

Working from the product and the container, we can establish the process parameters.

The principal parameters are:

1.   The process temperature
2.   The related processing time
3.   The pressure that can be applied without causing permanent container deformation

The prime parameters are supplemented by such secondary variables as:

(a)   The product filling temperature
(b)   The requisite product outlet temperature
(c)   The plant conditions (utilities)

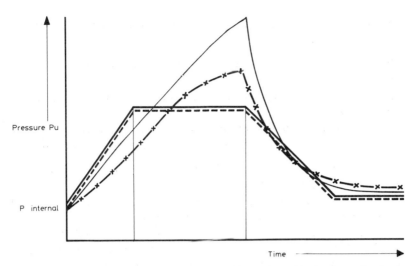

FIG. 10.   Temperature/pressure graph. The lines along the horizontal axis show the duration of the process and those along the vertical axis the pressure in bars. For medium Pu (thick continuous line), glass bottle (thin continuous line), can ($\times-\times-\times$), flexible container ($- - -$).

On the basis of the process parameters and the product to be handled, we can look at the sterilising equipment we should employ. Such equipment can be classified under two broad categories, namely equipment for batchwise sterilisation and equipment for automatic sterilisation.

7.   STERILISING SYSTEMS

Looking at an automatic sterilising system from the process parameter viewpoint, we can distinguish three key functions which the system must perform, namely:

Container infeed and discharge
In-steriliser container transport
Container processing (pressure, time and temperature)

In a batch sterilising system based on autoclaves these functions are:

Loading the autoclave crates
Taking the loaded crates to the autoclaves
Autoclaving the loaded crates by means of steam/air/water

In an automatic system, the following basic components are needed to carry out the key operations:

The facilities for automatic container supply and removal to and from the steriliser

The in-steriliser conveyor system that takes the containers through the sterilising zone

The processing system that provides automatic, load-dependent control of the main process parameters (time, temperature and pressure)

In batch sterilisation it is usual for the various functions and related operations to be performed manually.

Automatic sterilisation, on the other hand, is characterised by nonmanual container handling and process control.

Of course there are systems representing an intermediate between batchwise and automatic sterilisation. The simplest among them, introduced many years ago, is the vertical retort (see Fig. 11). The container-loaded crates are placed in the autoclave which is then closed and put into action, so that the containers undergo the requisite heat treatment.

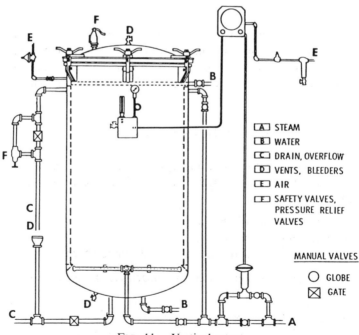

FIG. 11. Vertical retort.

This system has the following drawbacks:

High consumption of steam and water
High production costs as a result of manual loading, unloading and
process control
The autoclaving process is not uniform; manual control of individual
autoclaves may well lead to process variations

In automatic sterilisation, also known as continuous sterilisation, a
constant flow of containers moves from filling unit to steriliser. Here the
containers are transferred to the insteriliser conveyor system and taken
through the sterilising zone.

The continuous system most widely adopted is the hydrostatic system. It
has a sterilising zone that is in communication with the atmosphere through
columns or legs containing heads of water (see Fig. 12). The steam pressure
at the steam/water level is in balance with the hydrostatic pressure exerted
by the heads of water.

The containers are taken through the hydrostatic legs and the sterilising
zone in carriers attached to a conveyor chain.

A hydrostatic continuous steriliser's general configuration comprises

A preheating section
A sterilising zone
A cooling section

The unit's design characteristics depend upon the product (i.e. the required
processing time and temperature), the containers, the output needed, and
the plant conditions (utilities).

The key advantage offered by a continuous steriliser is that it makes each
container undergo the same uniform, reproducible process, and thus
greatly facilitates product data registration. This latter aspect is becoming
more and more significant, because food processors are required to keep
accurate records of the products sterilised in their plants.

The design of a continuous steriliser is established on the basis of various
data known as design criteria, which include:

The product
The container (plus closure)
The capacity
The number of units to be processed per annum
The plant utilities or consumables

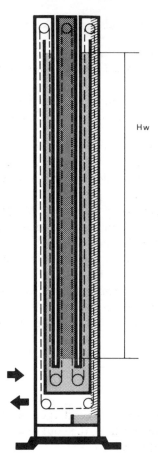

FIG. 12.   Hydrostatic principle. The system pressure corresponds to the water pressure exerted by the head Hw.

The product, or products, determine the shape of the carriers that are to ensure in-steriliser transport.

It is part of the steriliser manufacturer's specific capabilities to design a carrier that will take the packaged product(s) safely through the various processing sections.

The question whether a standard or non-standard system should be installed depends frequently on the variety of products and/or containers involved.

The container and its closure are equally important design criteria. As indicated earlier, the container's pressure resistance (P_W) plays a major rôle. Cans or glass bottles sealed with crown corks can be sterilised in saturated steam. In other words, adoption of the latter medium is possible when the container either has a very strong closure or can absorb a difference between internal and external pressure through deformation. Glass jars with a large opening (e.g. jars for infant food), plastic bottles and pouches do not possess this ability to resist pressure differences. If such containers are exposed to saturated steam, the resulting mechanical load on the seal will affect the seal's air-tightness, so that product sterility is no longer ensured.

To process the containers in question, an external excess pressure can be created by using a steam/air mixture or superheated water as a sterilising medium rather than saturated steam. This excess pressure will permit adequate sterilisation despite the container's low pressure resistance.

The pressure inside the container is built up by the factors mentioned earlier, i.e. $P\mathrm{vap}_{T_2}$, related to the sterilising temperature and P volume (air), related to the increase in volume of product and headspace.

The external pressure results from the air pressure and the water vapour pressure corresponding to the sterilising pressure. This can be illustrated most readily on the basis of a glass jar (Fig. 13).

For flexible containers, such as PE bottles and aluminium pouches application of the calculating procedure given in Fig. 13 is not warranted, because these containers have their own permissible deformation, so that in many cases it will be necessary to establish the overpressure by trial.

Frequently, overpressure is not only needed during sterilisation but also during the initial cooling phase. The external pressure induced by the saturated steam will fall off fairly rapidly, whilst the temperature inside the container is slow to drop. This causes the differential pressure to increase after sterilisation, so that cooling must be conducted at a given overpressure. The situation concerned can again be depicted by a graph (based on a hydrostatic system), as in Fig. 14. The graph clearly shows that $P2 > P1 > P3$.

As a third design criterion we have mentioned the capacity. In conjunction with the production schedule, indicating the amount of product to be processed per unit of time for a given period of time, an optimum must be established for the continuous steriliser. The first continuous sterilisers had a capacity of 125–150 1/1-litre cans per minute. Today, Stork Amsterdam builds sterilisers handling 1800 cans per minute, which corresponds to a capacity of 100 000 units per hour, or an annual production of 726×10^6 units.

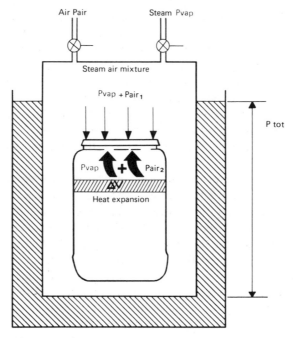

FIG. 13. Depending on the type of container (plus closure) selected, it is possible to establish what maximum pressure load will still ensure an air-tight seal. This pressure load, along with the filling temperature and the headspace, then allows the requisite over-pressure (Ptot $-$ Pvap) to be determined.

ΔV expansion $=$ hatched area.

        Pvap $=$ vapour pressure related to sterilising temperature. This pressure is the same inside and outside the container.

        Pair$_1$ $=$ external air pressure in steriliser.

        Pair$_2$ $=$ internal air pressure, formed by air temperature increase and volume increase through product expansion.

Finding the steriliser's optimum capacity is a complicated matter in which economic aspects play a dominant rôle. First of all the various design criteria must be reviewed and their interaction considered, whereupon the outcome of the study must be expressed in terms of system capacities.

   The utilities, or plant consumables, constitute a further design criterion. Those of interest in sterilisation are water, steam and the effluent resulting from the process. The latter especially is receiving more and more attention nowadays.

   In a sterilising system the product is heated from the filling temperature

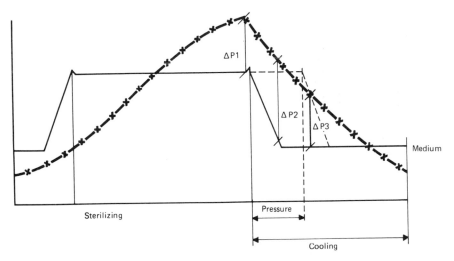

FIG. 14. Course of pressure in container and medium with and without pressure cooling. Horizontally, the time in minutes; vertically, the pressure in bars. Pw sterilisation ($\Delta P1$), Pw cooling ($\Delta P2$), Pw pressure cooling ($\Delta P3$), pressure cooling ($---$).

to the requisite sterilising temperature and then cooled down again to the outlet temperature. The process involved requires a comparatively high energy input for preheating and a large amount of water for cooling purposes. However, in a Stork continuous steriliser (named Hydromatic), the consumption of steam and water is drastically reduced by regeneration. From the outgoing hydrostatic leg, which picks up heat from the sterilised containers, hot water is pumped to the preheating leg. The water in the latter leg imparts its heat to the incoming containers, whose temperature thus becomes higher than the filling temperature before they enter the sterilising zone. This regenerating system makes it possible to save up to 60 or 70% on steam.

The consumption of cooling water can be reduced considerably through the use of recirculation. It involves the creation by means of heat exchangers of a primary cooling circuit whose water circulates through the sterilising system and cools down the containers.

Next to this, the system has a secondary circuit which is only subject to thermal load and which either contains surface water or water intended for use elsewhere as boiler feed water, process water, product cleaning water, etc. after it has served its cooling function.

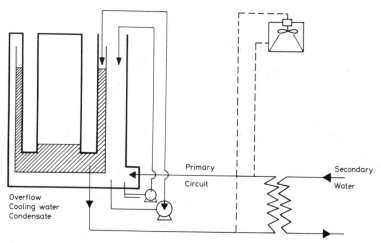

FIG. 15.   Continuous steriliser with closed-circuit cooling system.

Sometimes, open cooling towers are employed to discharge the heat absorbed, in which case a minimal amount of make-up water will be needed (see Fig. 15).

8.   CONCLUSIONS

Figure 16 shows a schematic survey of the factors that play a rôle in thermal food preservation.

Aside from the factors shown in the overview, there is a range of economic aspects underlying the choice of sterilising system. Weighing up these factors and aspects can lead to a decision on pragmatic grounds. It is also possible, however, to adopt a different basis in approaching the selection problem, namely that of risk analysis.

Each sterilising system involves a certain risk potential in that the product may be understerilised or reinfected after sterilisation in the final container. Such understerilisation or reinfection can ensue from:

Overfill
Operator handling errors
System failure
Faulty product routing, which results in a mix-up between sterilised and non-sterilised product.

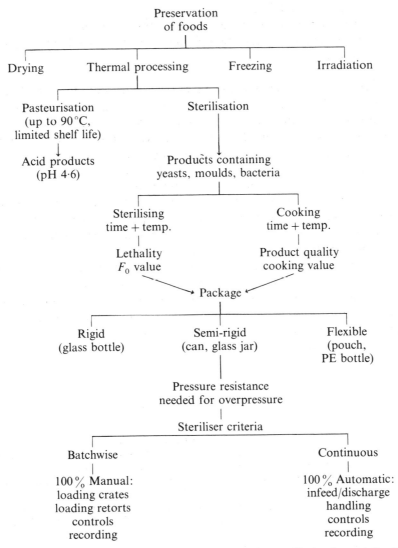

FIG. 16. A schematic survey of the factors that play a rôle in thermal food preservation.

Container damage caused by:

rough can handling on the conveyors
improper conveyor layout, so that containers may become jammed
rough container handling in loading and/or unloading the sterilising
system

Expressing the outcome of a risk analysis in terms of money has a major effect on the economic model presented by capital cost projections. Substandard product quality through malperformance is bound to put production figures, market shares and supply contracts under considerable pressure, which implies that the risks run by food processors are serious and that proper control of these risks is a vital issue.

As regards the future, the following trends are discernible in thermal food preservation. First of all, there is a growing demand for better product quality. This may result in process margins becoming narrower through the adoption of higher temperatures and shorter times, which entails the risk of product spoilage by inadequate enzyme inactivation.

In the packaging industry we see the emergence of double-reduced tinplate, the two-piece deep-drawn can and the welded can. The latest development in this area is the tin-free, lacquered can whose application has its origin in declining tin resources. The trends in question should make food processors stop and think hard about where to go from here, because their trade is one that affects the sustenance of billions.

REFERENCES

1. GUYER, R. B. and HOLMQUIST, J. W., *Enzyme Regeneration in High Temperature Short Time Sterilized Canned Foods*, Continental Can Company, USA, 1954.
2. YAMAGUCHI, K. and KISHIMOTO, A., *In-Package HTST-Sterilization of Foods Packaged in Retortable Pouches*, Toyo Seikan/Toyo Kohan Company R&D Centre, Yokohama, 1974.
3. REICHERT, J. E., Der C-Welt als Hilfmittel zur Process optimierung, *Zeitschrift für Lebensmittel Technologie und Verfahrens-Technik*, 1972, **28**(1/77), 1–7.

The following books are recommended for use as works of reference:

National Canners Association. *Canned Foods. Principles of Thermal Process Control and Container Closure Evaluation*. Distributed by The Food Processors Institute, 1950 Sixth Street, Berkeley, California 94710, USA.

STUMBO, C. R., *Thermobacteriology in Food Processing*, Academic Press, New York, 1973.

DOWNEY, W. K. (Ed.), *Food Quality and Nutrition*, Applied Science Publishers, London, 1978.

HERMANS, W. F. (Stork Research Centre, Amsterdam), *Continuous Feed Processing Flexibles*, Chipping Campden Food Preservation Research Station, UK, 1979.

INDEX